Springer Series in Information Sciences 18

Editor: Thomas S. Huang

Springer Series in Information Sciences

Editors: Thomas S. Huang   Teuvo Kohonen   Manfred R. Schroeder

Managing Editor: H. K. V. Lotsch

C. K. Chui   G. Chen

Linear Systems and Optimal Control

With 4 Figures

Springer-Verlag Berlin Heidelberg GmbH

Professor Dr. Charles K. Chui

Department of Mathematics and Department of Electrical Engineering,
Texas A & M University, College Station, TX 77843, USA

Dr. Guanrong Chen

Department of Electrical and Computer Engineering,
Rice University, Houston, TX 77251, USA

Series Editors:

Professor Thomas S. Huang

Department of Electrical Engineering and Coordinated Science Laboratory,
University of Illinois, Urbana, IL 61801, USA

Professor Teuvo Kohonen

Department of Technical Physics, Helsinki University of Technology,
SF-02150 Espoo 15, Finland

Professor Dr. Manfred R. Schroeder

Drittes Physikalisches Institut, Universität Göttingen, Bürgerstrasse 42–44,
D-3400 Göttingen, Fed. Rep. of Germany

Managing Editor: Helmut K. V. Lotsch

Springer-Verlag, Tiergartenstrasse 17,
D-6900 Heidelberg, Fed. Rep. of Germany

ISBN 978-3-642-64787-1   ISBN 978-3-642-61312-8 (eBook)
DOI 10.1007/978-3-642-61312-8

Library of Congress Cataloging-in-Publication Data. Chui, C.K. Linear systems and optimal control.
(Springer series in information sciences ; 18) Bibliography: p. Includes index. 1. Control theory. 2. Optimal
control. I. Chen, G. (Guanrong) II. Title. III. Series. QA402.3.C5566 1988   629.8'312   88-2012

© Springer-Verlag Berlin Heidelberg 1989
Originally published by Springer-Verlag Berlin Heidelberg New York in 1989
Softcover reprint of the hardcover 1st edition 1989

Typesetting: Macmillan India Ltd., India

2154/3150-543210 – Printed on acid-free paper

Preface

A knowledge of linear systems provides a firm foundation for the study of optimal control theory and many areas of system theory and signal processing. State-space techniques developed since the early sixties have been proved to be very effective. The main objective of this book is to present a brief and somewhat complete investigation on the theory of linear systems, with emphasis on these techniques, in both continuous-time and discrete-time settings, and to demonstrate an application to the study of elementary (linear and nonlinear) optimal control theory.

An essential feature of the state-space approach is that both time-varying and time-invariant systems are treated systematically. When time-varying systems are considered, another important subject that depends very much on the state-space formulation is perhaps real-time filtering, prediction, and smoothing via the Kalman filter. This subject is treated in our monograph entitled "Kalman Filtering with Real-Time Applications" published in this Springer Series in Information Sciences (Volume 17). For time-invariant systems, the recent frequency domain approaches using the techniques of Adamjan, Arov, and Krein (also known as AAK), balanced realization, and H^{∞} theory via Nevanlinna-Pick interpolation seem very promising, and this will be studied in our forthcoming monograph entitled "Mathematical Approach to Signal Processing and System Theory". The present elementary treatise on linear system theory should provide enough engineering and mathematics background and motivation for study of these two subjects.

Although the style of writing in this book is intended to be informal, the mathematical argument throughout is rigorous. In addition, this book is self-contained, elementary, and easily readable by anyone, student or professional, with a minimal knowledge of linear algebra and ordinary differential equations. Most of the fundamental topics in linear systems and optimal control theory are treated carefully, first in continuous-time and then in discrete-time settings. Other related topics are briefly discussed in the chapter entitled "Notes and References". Each of the six chapters on linear systems and the three chapters on optimal control contains a variety of exercises for the purpose of illustrating certain related view-points, improving the understanding of the material, or filling in the details of some proofs in the text. For this reason, the reader is encouraged to work on these problems and refer to the "answers and hints" which are included at the end of the text if any difficulty should arise.

This book is designed to serve two purposes: it is written not only for self-study but also for use in a one-quarter or one-semester introductory course in linear systems and control theory for upper-division undergraduate or first-year graduate engineering and mathematics students. Some of the chapters may be covered in one week and others in at most two weeks. For a fifteen-week semester, the instructor may also wish to spend a couple of weeks on the topics discussed in the "Notes and References" section, using the cited articles as supplementary material.

The authors are indebted to Susan Trussell for typing the manuscript and are very grateful to their families for their patience and understanding.

College Station   *Charles K. Chui*
Texas, May 1988   *Guanrong Chen*

Contents

1. State-Space Descriptions

Although the history of linear system theory can be traced back to the last century, the so-called state-space approach was not available till the early 1960s. An important feature of this approach over the traditional frequency domain considerations is that both time-varying and time-invariant linear or nonlinear systems can be treated systematically. The purpose of this chapter is to introduce the state-space concept.

1.1 Introduction

A typical model that applied mathematicians and system engineers consider is a "machine" with an "input-output" relation placed at the two terminals (Fig. 1.1). This machine is also called a *system* which may represent certain biological, economical, or physical systems, or a mathematical description in terms of an algorithm, a system of integral or differential equations, etc. In many applications, a system is described by the totality of input-output relations (u, v) where u and v are functions or, when discretized, sequences, and may be either scalar or vector-valued. It should be emphasized that the collection of all input-output ordered pairs is not necessarily single-valued. As a simple example, consider a system given by the differential equation $v'' + v = u$. In this situation, the totality of all input-output relations that determines the system is the set

$$S = \{(u, v): v'' + v = u\}$$

and it is clear that the same input u gives rise to infinitely many outputs v. For example, $(1, \sin t + 1)$, $(1, \cos t + 1)$, and even $(1, a\cos t + b\sin t + 1)$ for arbitrary constants a and b, all belong to S. To avoid such an unpleasant situation and to give a more descriptive representation of the system, the "state" of the system is considered. The *state* of a system explains its past, present, and future situations. This is done by introducing a minimum number of variables which are called *state variables* that represent the present situation, using the past information, namely the initial state, and describe the future behavior of the system completely. The column vector of the state variables, in a given order, is called a *state vector*.

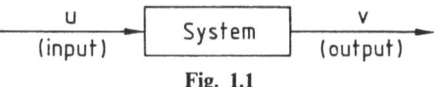

Fig. 1.1

Let us return to the simple example of the system described by the differential equation $v'' + v = u$ with a specified initial state. Introducing the state vector

$$x = \begin{bmatrix} x_1 \\ x_2 \end{bmatrix},$$

where x_1 and x_2 are state variables satisfying the initial state $x_1(a) = b$ and $x_2(a) = c$, we can give a "state-space" description of this system by using a system of two equations:

$$\dot{x} = \begin{bmatrix} 0 & 1 \\ -1 & 0 \end{bmatrix} x + \begin{bmatrix} 0 \\ 1 \end{bmatrix} u$$

$$v = [1 \ 0]x \ , \tag{1.1}$$

where $\dot{x}$ denotes the derivative of the state vector x. The definition of *state-space* will be better understood later in Sect. 1.4. Here, the first equation in (1.1) gives the input-state relation while the second equation describes the state-output relation. The so-called *state-space equations* (1.1) could be obtained by setting the state variables x_1 and x_2 to be v and v' respectively. However, without the knowledge of such substitutions, it may not be immediately clear that the input-output relation follows from the state-space equations (1.1). To demonstrate how this is done more generally, we rewrite (1.1) as

$$\dot{x} = Ax + Bu$$

$$v = Cx \tag{1.2}$$

where A, B, C are 2×2, 2×1, 1×2 matrices and let $p(\lambda)$ be the characteristic polynomial of A. In this example, $p(\lambda) = \lambda^2 + 1$, so that by the Cayley–Hamilton Theorem, we have

$$p(A) = A^2 + I = 0 \ .$$

Hence, differentiating the second equation in (1.2) twice (the number of times of differentiation will equal the degree of the characteristic polynomial of the square matrix A), and utilizing the first equation in (1.2) repeatedly, we have

$$Cx = v$$

$$CAx = v' - CBu$$

$$CA^2x = v'' - CBu' - CABu \ .$$

Therefore, the identity $p(A) = A^2 + I = 0$ can be used to eliminate x, yielding:

$$(v'' - CBu' - CABu) + v = CA^2 x + Cx = C(A^2 + I)x = 0 \quad \text{or}$$

$$v'' + v = C(Bu' + ABu)$$

$$= [1 \quad 0] \left(\begin{bmatrix} 0 \\ 1 \end{bmatrix} u' + \begin{bmatrix} 1 \\ 0 \end{bmatrix} u \right)$$

$$= u \ .$$

1.2 An Example of Input-Output Relations

More generally, if the characteristic polynomial of an $n \times n$ matrix A in an input-state equation such as (1.2) is

$$p(\lambda) = \lambda^n + a_1 \lambda^{n-1} + \ldots + a_n \ ,$$

then the above procedure gives

$$Cx = v$$

$$CAx = v' - CBu$$

$$CA^2 x = v'' - CBu' - CABu$$

$$\ldots$$

$$CA^n x = v^{(n)} - CBu^{(n-1)} - CABu^{(n-2)} - \ldots - CA^{n-1}Bu \ ,$$

so that, by setting $a_0 = 1$, we have:

$$\sum_{k=0}^{n} a_k \left(v^{(n-k)} - C \sum_{j=0}^{n-k-1} A^j Bu^{(n-k-j-1)} \right) = Cp(A)x = 0 \ .$$

That is, the input-output relation can be given by

$$\sum_{j=0}^{n} a_j v^{(n-j)} = C \sum_{k=0}^{n} a_k \sum_{j=0}^{n-k-1} A^j Bu^{(n-k-j-1)} \tag{1.3}$$

with $a_0 = 1$.

A slightly more general form of (1.3) is given by

$$Lv = Mu$$

$$L = \sum_{j=0}^{n} a_j \frac{d^{n-j}}{dt^{n-j}} \ , \quad a_0 = 1 \tag{1.4}$$

$$M = \sum_{k=0}^{m} b_k \frac{d^{m-k}}{dt^{m-k}} \ , \quad m \le n \ .$$

However, the system with input-output relations described by (1.4) does not necessarily have a state-space description given by (1.2) (Exercise 1.2). We also remark in passing that even if it has such a description, the matrices A, B and C are not unique (Exercise 1.3).

1.3 An Example of State-Space Descriptions

A more general state-space description of a system with input-output pairs (u, v) is given by

$$\dot{x} = Ax + Bu$$
$$v = Cx + Du$$

(1.5)

where A, B, C, D are matrices with appropriate dimensions. By eliminating the state vector x and its derivative with the help of the Cayley-Hamilton Theorem as above, it is not difficult to see that the input-output pair (u, v) in (1.5) satisfies the relation $Lv = Mu$ in (1.4) with appropriate choices of constants a_j and b_k (Exercise 1.4). To see the converse, that is, to show that the input-output relations in (1.4) have a state-space description as given in (1.5), we follow the standard technique of transforming an nth order linear differential equation to a first order vector differential equation as was done in the simple example discussed earlier by choosing the matrix A to be

$$\begin{bmatrix} 0 & 1 & 0 & \cdots & 0 \\ 0 & 0 & 1 & & \vdots \\ \vdots & & \ddots & \ddots & 0 \\ 0 & \cdots & & 0 & 1 \\ -a_n & \cdots & & -a_2 & -a_1 \end{bmatrix} .$$

Of course there are other choices of A. But with this "so-called" standard choice, it is clear that the matrix C must be given by

$$C = [1 \ 0 \ldots 0] .$$

Hence, by setting $B = [\beta_1 \ldots \beta_n]^T$ and $D = [\beta_0]$ we see that the variables of the vector $x = [x_1 \ldots x_n]^T$ in (1.5) satisfy the equations:

$$x_1' = x_2 + \beta_1 u$$
$$x_2' = x_3 + \beta_2 u$$
$$\cdots$$
$$x_{n-1}' = x_n + \beta_{n-1} u$$
$$x_n' + a_1 x_n + \ldots + a_n x_1 = \beta_n u$$
$$v = x_1 + \beta_0 u .$$

That is, the state variables are defined by

$$x_1 = v - \beta_0 u$$

$$x_2 = x_1' - \beta_1 u = v' - (\beta_0 u' + \beta_1 u)$$

$$x_3 = x_2' - \beta_2 u = v'' - (\beta_0 u'' + \beta_1 u' + \beta_2 u)$$

. . .

$$x_n = x_{n-1}' - \beta_{n-1} u = v^{(n-1)} - (\beta_0 u^{(n-1)} + \ldots + \beta_{n-1} u)$$

and must satisfy the constraint:

$$x_n' + a_1 x_n + \ldots + a_n x_1 = \beta_n u \, ,$$

or equivalently,

$$\sum_{j=0}^{n} a_j v^{(n-j)} = \left(\sum_{i=0}^{n} a_i \beta_{n-i} \right) u + \left(\sum_{i=0}^{n-1} a_i \beta_{n-i-1} \right) u'$$

$$+ \ldots + (a_1 \beta_0 + a_0 \beta_1) u^{(n-1)} + a_0 \beta_0 u^{(n)} \, . \tag{1.6}$$

Hence, the constants $\beta_0, \ldots, \beta_n$ are uniquely determined by the linear matrix equation

$$\begin{bmatrix} a_0 & a_1 & \cdots & a_n \\ 0 & & & \vdots \\ \vdots & & & a_1 \\ 0 & \cdots & 0 & a_0 \end{bmatrix} \begin{bmatrix} \beta_n \\ \vdots \\ \beta_0 \end{bmatrix} = \begin{bmatrix} b_m \\ \vdots \\ b_{m-n} \end{bmatrix}$$

where $a_0 = 1$ and $b_j = 0$ for $j < 0$. We remark that the highest derivative of u in (1.6) is n, and hence the order m of the differential operator M in (1.4) is not allowed to exceed n.

1.4 State-Space Models

A system with the state-space description given by (1.5) is usually called a single-input/single-output *time-invariant* system; that is, the matrices A, B, C and D in (1.5) are constant matrices and the input and output functions are scalar-valued. In general, we have to work with *time-varying* systems, and in addition, the input and output functions may happen to be vector-valued; in other words, we may have a multi-input/multi-output system. The state-space description of such a system is given by

$$\dot{x} = A(t)x + B(t)u$$
$$v = C(t)x + D(t)u \, . \tag{1.7}$$

The digital version of (1.7) is

$$x_{k+1} = A_k x_k + B_k u_k$$
$$v_k = C_k x_k + D_k u_k \ , \tag{1.8}$$

where $\{u_k\}$ and $\{v_k\}$ are input and output sequences of the discretized (or digital) system, respectively. Of course (1.8) is only an approximation of (1.7), for instance, by setting $u_k = u(kh)$, $v_k = v(kh)$, and $x_k = x(kh)$ where h is a sampling time unit. A natural choice of the matrices A_k, B_k, C_k and D_k is given by

$$A_k = hA(kh) + I$$

$$B_k = B(kh)$$

$$C_k = C(kh) \quad \text{and}$$

$$D_k = D(kh) \ .$$

A small sampling time unit is necessary to give a good approximation. We will be dealing with the state-space descriptions (1.7, 8) for continuous-time and discrete-time systems, respectively. The vector space, spanned by the state vectors which are generated by all "admissible" inputs and initial states, is called the *state-space*. For a better understanding, see Exercises 2.2–4.

It will be clear from Exercise 2.5 that the outputs in the state-space descriptions (1.7, 8) are linear in the state vectors for zero input and linear in the inputs for zero initial state. For this reason, the systems we consider here are called *linear systems*. In the subject of *control theory*, linear systems are also called *linear dynamic systems*, the state-space descriptions (1.7, 8), *dynamic equations*, and the matrices $A(t)$, $B(t)$, $C(t)$, and $D(t)$ in (1.7) or A_k, B_k, C_k, and D_k in (1.8) are called *system* (or *dynamic*), *control*, *observation* (or *output*), and *transfer matrices*, respectively.

Exercises

1.1 Give a state-space description for the input-output relations $v'' + av' + bv = u$ by using the state variables $x_1 = \alpha v + \beta v'$ and $x_2 = \gamma v + \delta v'$ where $\alpha\delta - \beta\gamma \neq 0$.

1.2 Determine all constants a, b and c so that the linear system with input-output relations $v'' + v' = au + bu' + cu''$ has a state-space description of the form given by (1.2).

1.3 By using Exercise 1.1, show that the matrices A, B, and C in the state-space description (1.2) for the linear system with input-output relations $v'' + av' + bv = 0$ are not unique.

1.4 Determine the constants a_j and b_k in (1.4) for the input-output relations of the linear system (1.5) where A, B, C and D are arbitrary $n \times n$, $n \times 1$, $1 \times n$, and 1×1 matrices.

1.5 (a) Give a state-space description for the two-input and two-output system

$$v_1'' + a_{11}v_1' + a_{12}v_1 + b_{11}v_2' + b_{12}v_2 = \alpha_1 u_1 + \beta_1 u_2$$

$$v_2'' + a_{21}v_1' + a_{22}v_1 + b_{21}v_2' + b_{22}v_2 = \alpha_2 u_1 + \beta_2 u_2 \ .$$

(b) Derive a general state-space description for the normal n-input and n-output system

$$v_1^{(n)} + \sum_{j=1}^{n} \{a_{1j}^1 v_1^{(n-j)} + a_{1j}^2 v_2^{(n-j)} + \ \ldots \ + a_{1j}^n v_n^{(n-j)}\} = \sum_{j=1}^{n} \alpha_{1j} u_j \ .$$

$$\ldots$$

$$v_n^{(n)} + \sum_{j=1}^{n} \{a_{nj}^1 v_1^{(n-j)} + a_{nj}^2 v_2^{(n-j)} + \ \ldots \ + a_{nj}^n v_n^{(n-j)}\} = \sum_{j=1}^{n} \alpha_{nj} u_j \ .$$

1.6 (a) Give a state-space description for the discrete-time system defined by the difference equation

$$v_{k+2} + v_{k+1} + v_k = u_k \ .$$

$$\left(\text{Hint: Let } x_{1,k} = v_k, x_{2,k} = v_{k+1} \text{ and} \right.$$

$$\left. x_k = \begin{bmatrix} x_{1,k} \\ x_{2,k} \end{bmatrix} \right).$$

(b) Derive a general state-space description for the discrete-time system defined by the difference equation

$$a_0 v_{k+n} + a_1 v_{k+n-1} + \ \ldots \ + a_n v_k = b_0 u_{k+m} + \ \ldots \ + b_m u_k \ ,$$

where $a_0 = 1$, $m \leq n$, and m, n are arbitrary positive integers.

2. State Transition Equations and Matrices

In this chapter, we will discuss the solution of the state-space equation assuming that the initial state as well as all the governing matrices are given. Both continuous-time and discrete-time systems will be considered. It is clear that only the input-state equation has to be solved.

2.1 Continuous-Time Linear Systems

From the theory of ordinary differential equations, if $A(t)$ is an $n \times n$ matrix whose entries are continuous functions on an interval J which contains t_0 in its interior, then the initial value problem

$$\dot{x} = A(t)x$$
$$x(t_0) = e_i \ , \tag{2.1}$$

where $e_i = [0 \ldots 0 \ 1 \ 0 \ldots 0]^T$, the entry 1 being the ith component, has a unique solution which we will denote by $\phi_i(t, t_0)$. Let $\Phi(t, t_0)$ be the $n \times n$ matrix with $\phi_i(t, t_0)$ as its ith column. Since these column vectors are linearly independent, the "fundamental matrix" $\Phi(t, t_0)$ is nonsingular. For convenience, we assume that J is an open interval. Since the above discussion is valid for any t_0 in J, we could consider $\Phi(s, t)$ as a matrix-valued function of two variables in J. Clearly,

$$\Phi(t, t) = I \ ,$$

the identity matrix, for all t in J. Set

$$F(s, t) = \Phi(s, \tau)\Phi^{-1}(t, \tau) \ .$$

Then $F(s, \tau) = \Phi(s, \tau)\Phi^{-1}(\tau, \tau) = \Phi(s, \tau)$, i.e., $F \equiv \Phi$, so that

$$\Phi(s, t) = \Phi(s, \tau)\Phi^{-1}(t, \tau)$$

or, equivalently, $\Phi(s, t)$ satisfies the "transition" property:

$$\Phi(s, \tau) = \Phi(s, t)\Phi(t, \tau) \ , \tag{2.2}$$

where s, t, and τ are in J.

We now consider the input-state equation with a given initial state x_0 at time t_0, namely

$$\dot{x} = A(t)x + B(t)u$$
$$x(t_0) = x_0 \ , \tag{2.3}$$

where $A(t)$ and $B(t)$ are $n \times n$ and $n \times p$ matrices respectively, and u is a p-dimensional column vector. Although weaker conditions are allowed, we will always assume, for convenience, that all entries of $A(t)$ are continuous functions on J and that the entries of $B(t)$ as well as the components of u are piecewise continuous on J. Again from the theory of ordinary differential equations, (2.3) has a unique solution given by

$$x(t) = \Phi(t, t_0)x(t_0) + \int_{t_0}^{t} \Phi(t, \tau)B(\tau)u(\tau)d\tau \ , \tag{2.4}$$

where, as usual, integration is performed componentwise, and $\Phi(t, t_0)$ is the fundamental matrix of the first order homogeneous equation $\dot{x} = Ax$ discussed above. In the subject of control theory, one could think of u as the control function that takes an initial state $x(t_0)$ to a state $x(t)$ in continuous time from time t_0 to time t, and "equation" (2.4) describes how this is done. Because of its formulation, this equation is also called the (continuous-time) integral equation of u. Note that the solution of this equation for the control function u that takes $x(t_0)$ to $x(t)$ is given by the input-state equation (2.3). The matrix $\Phi(t, t_0)$ that describes this transition process is usually called the *transition matrix* of the linear system.

2.2 Picard's Iteration

In order to have a better understanding of the transition process, it is important to study the transition matrix. We first consider the special case where $A = [a_{ij}]$ is a constant matrix. Denote by $|A|_1$ the l^1 norm of this matrix; that is

$$|A|_1 = \sum_{i, j} |a_{ij}| \ .$$

By Exercise 2.8, we have $|A^2|_1 \leq |A|_1^2, \ldots, |A^n|_1 \leq |A|_1^n, \ldots$, and this allows us to define

$$e^{tA} = \sum_{n=0}^{\infty} \frac{t^n}{n!} A^n$$

since the sequence of partial sums of the infinite series is a Cauchy sequence:

$$\left| \sum_{n=M}^{N} \frac{t^n}{n!} A^n \right|_1 \leq \sum_{n=M}^{N} \frac{|t|^n}{n!} |A^n|_1$$

$$\leq \sum_{n=M}^{N} \frac{(|t||A|_1)^n}{n!}$$

which tends to 0 as M and N tend to infinity independently. (Here, the triangle inequality in Exercise 2.8 has been used.) In addition, it is also clear from this infinite series definition that

$$\frac{d}{dt} e^{tA} = A e^{tA} \ .$$

Hence, it follows immediately that the solution $\phi_i(t, t_0)$ of (2.1) is given by

$$\phi_i(t, t_0) = e^{(t-t_0)A} e_i \ ;$$

that is, the transition matrix in (2.4) for the system with constant system matrix A is given by

$$\Phi(t, t_0) = e^{(t-t_0)A} \ . \tag{2.5}$$

When $A = A(t)$ is not a constant, that is when time-varying state-space equations are considered, an explicit formulation of the transition matrix is usually difficult to obtain. The following iteration process, usually attributed to Picard, gives an approximation of $\Phi(t, t_0)$. Again, for convenience, we assume that the entries of $A(t)$ are bounded functions in J, so that a positive constant C exists with

$$|A(t)|_1 \leq C < \infty, \quad t \in J \ .$$

We start with the identity matrix. Set

$$P_0(t) = I$$

$$P_1(t) = I + \int_{t_0}^{t} A(s) P_0(s) ds$$

. . .

$$P_N(t) = I + \int_{t_0}^{t} A(s) P_{N-1}(s) ds \ .$$

Then for all $t \in J$ and $N > M$, we have

$$|P_N(t) - P_M(t)|_1 = \left| \sum_{k=M}^{N-1} [P_{k+1}(t) - P_k(t)] \right|_1$$

$$= \left| \sum_{k=M}^{N-1} \int_{t_0}^{t} A(s_1) \int_{t_0}^{s_1} A(s_2) \ldots \int_{t_0}^{s_k} A(s_{k+1}) ds_{k+1} \ldots ds_1 \right|_1$$

$$\leq \sum_{k=M}^{N-1} \left| \int_{t_0}^{t} \ldots \int_{t_0}^{s_k} ds_{k+1} \ldots ds_1 \right| C^{k+1}$$

$$= \sum_{k=M}^{N-1} \frac{(C|t-t_0|)^{k+1}}{(k+1)!}$$

which tends to zero uniformly on any bounded interval as $M, N \to \infty$ independently. That is, $\{P_N(t)\}$ is a Cauchy sequence of matrix-valued continuously differentiable functions on J. Let $P(t, t_0)$ be its uniform limit. Since

$$\frac{d}{dt} P_N(t) = A(t)P_{N-1}(t)$$

and $P_N(t_0) = I$, it follows from a theorem of Weierstrass that

$$\frac{d}{dt} P(t, t_0) = A(t)P(t, t_0)$$

$$P(t_0, t_0) = I .$$

This, of course, means that the columns of $P(t, t_0)$ are the unique solutions $\phi_i(t, t_0)$ of the initial value input-state equations (2.1), so that $P(t, t_0)$ coincides with $\Phi(t, t_0)$. We have now described a simple iteration process that gives a uniform approximation of $\Phi(t, t_0)$. It also allows us to write:

$$\Phi(t, t_0) = I + \int_{t_0}^{t} A(s)ds + \int_{t_0}^{t} A(s_1) \int_{t_0}^{s_1} A(s_2)ds_2 ds_1 + \ldots . \tag{2.6}$$

It is clear that if $A = A(t)$ is a constant matrix, then (2.5) and (2.6) are identical, using the definition of $\exp[(t-t_0)A]$.

2.3 Discrete-Time Linear Systems

We now turn to the discrete-time system. The input-state equation with a given initial state x_0 is given by

$$x_{k+1} = A_k x_k + B_k u_k, \quad k = 0, 1, \ldots, \tag{2.7}$$

where A_k and B_k are $n \times n$ and $n \times p$ matrices and u_k, $k = 0, 1, \ldots$, are p-dimensional column vectors. Writing out (2.7) for $k = 0, 1, \ldots$, respectively, we have

$$x_1 = A_0 x_0 + B_0 u_0$$

$$x_2 = A_1 x_1 + B_1 u_1$$

$$\ldots$$

$$x_{k+1} = A_k x_k + B_k u_k .$$

Hence, by substituting the first equation into the second one, and this new equation into the third one, etc., we obtain

$$x_N = \Phi_{N0} x_0 + \sum_{k=1}^{N} \Phi_{Nk} B_{k-1} u_{k-1} \tag{2.8}$$

where we have defined the "transition" matrices:

$$\Phi_{kk} = I$$
$$\Phi_{jk} = A_{j-1} \ldots A_k \quad \text{for} \quad j > k . \tag{2.9}$$

In particular, if $A_k = A$ for all k, then $\Phi_{jk} = A^{j-k}$ for $j \geq k$. Equation (2.8) is called the (discrete-time) *state transition equation* corresponding to the input-state equation (2.7) and Φ_{jk} $(j \geq k)$ are called the *transition matrices*. The state transition equation describes the transition rule in discrete-time that the control sequence $\{u_k\}$ takes the initial state x_0 to the final state x_N. We remark, however, that although the transition matrices Φ_{jk} satisfy the "transition" property

$$\Phi_{ik} = \Phi_{ij} \Phi_{jk} \quad \text{for} \quad i \geq j \geq k ,$$

Φ_{jk} is *not defined* for $j < k$, and in fact, even if $A_k = A$ for all k, Φ_{ik} $(i > k)$ is singular if A is. This shows that discrete-time and continuous-time linear systems may have different behaviors. However, if the system matrices $A_k, \ldots, A_{j-1}$, where $k < j$, are nonsingular, it is natural to introduce the notation $\Phi_{kj} = A_k^{-1} \ldots A_{j-1}^{-1}$, so that $\Phi_{kj} = \Phi_{jk}^{-1}$ or $\Phi_{kj} \Phi_{jk} = I$, completing the transition property.

2.4 Discretization

If the discrete-time state-space description

$$x_{k+1} = A_k x_k + B_k u_k$$
$$v_k = C_k x_k + D_k u_k \qquad\qquad (2.10)$$

is obtained as an approximation of the continuous-time state-space description

$$\dot{x} = A(t)x + B(t)u$$
$$v = C(t)x + D(t)u \qquad\qquad (2.11)$$

by setting, say, $x_k = x(kh)$, $u_k = u(kh)$ and $v_k = v(kh)$, then the singularity of the matrices A_k, and consequently of the transition matrices Φ_{jk} ($j \geq k$), may result from applying a poor discretization method. In order to illustrate our point here, we only consider the case where $A = A(t)$ is a constant matrix.

As pointed out in the last chapter, a "natural" choice of A_k is

$$A_k = I + hA(kh) , \qquad\qquad (2.12)$$

the reason being

$$x_{k+1} - x_k \doteq h\dot{x}(kh) \doteq h(A(kh)x_k - B(kh)u_k) .$$

Of course, if the time sample h is very small then A_k will usually be nonsingular. However, in many applications, some entries of A may be very large negative numbers so that it would become difficult, and sometimes even numerically unstable, to choose very small h. The state transition equation (2.4) with the transition matrix given in (2.5), being an integral equation, gives a much more numerically stable discretization. Setting $t_0 = kh$ and $t = (k+1)h$, we have

$$x_{k+1} \doteq \Phi((k+1)h, kh)x_k + \int_{kh}^{(k+1)h} \Phi((k+1)h, \tau) B(\tau)u(\tau)d\tau , \qquad (2.13)$$

so that the matrix A_k in the discrete-time state-space description (2.10) is now

$$A_k = \Phi((k+1)h, kh) . \qquad\qquad (2.14)$$

This is a nonsingular matrix, and consequently the corresponding transition matrix becomes

$$\Phi_{ij} = \Phi(ih, jh) .$$

We note, in particular, that the restriction $i \geq j$ can now be removed. We also remark that if A is a constant matrix the choice of A in (2.12) as a result of discretizing the input-state equation (2.11) gives only the linear term in the series definition of $\exp(hA)$. To complete the discretization procedure in (2.13), we could replace $u(\tau)$ by u_k and apply any simple integration quadrature to the remaining integral. If, for instance, both A and B in the continuous-time state-space description (2.11) are constant matrices, then the remaining integral is precisely

$$\int_0^h e^{tA}\,dt = h\left(I + \frac{h}{2!}A + \frac{h^2}{3!}A^2 + \dots\right)$$

and the matrix B_k in the corresponding discrete-time state-space (2.10) description becomes

$$B_k = h\left(I + \frac{h}{2!}A + \frac{h^2}{3!}A^2 + \dots\right)B$$

which is again a constant matrix.

Exercises

2.1 Solve the differential equation (2.1) for

$$A = \begin{bmatrix} 1 & t \\ 0 & 1 \end{bmatrix}$$

and determine the corresponding transition matrix $\Phi(t, t_0)$.

2.2 Recall that the state space X is the vector space of all (vector-valued) functions each of which is a (unique) solution of (2.3) for some initial state and some input (or control) u. Consider an admissible class $\mathscr{U}$ of input functions and let $X(\mathscr{U})$ be the subspace of X where only input functions in $\mathscr{U}$ are used. Determine $X(\mathscr{U})$ for $A = [0]$, $B = [1]$ and $\mathscr{U} = \mathrm{sp}\{1, \dots, t^N\}$, the linear span of $1, \dots, t^N$.

2.3 Repeat Exercise 2.2 for the admissible class $\mathscr{U} = \mathrm{sp}\{u_0, \dots, u_N\}$ where

$$u_i(t) = \begin{cases} 0 & \text{if } t < t_i \\ 1 & \text{if } t \geq t_i \end{cases}$$

and $0 = t_0 < t_1 < \dots < t_N < \infty$.

2.4 Refer to Exercise 2.2 for the necessary definitions. Let

$$A = \begin{bmatrix} 0 & 1 \\ 0 & 0 \end{bmatrix} \quad \text{and} \quad B = \begin{bmatrix} 1 \\ t \end{bmatrix}.$$

Find a basis of $X(\mathrm{sp}\{1, \dots, t^N\})$.
(*Hint*: Use the state transition equation.)

2.5 Show that a system with state-space description given by (2.10) or (2.11) is indeed a linear system in the sense that the output is linear in the state vectors for zero input and linear in the input vectors for zero initial state. (*Hint*: An operator L is said to be linear if $L(ay+bz)=aLy+bLz$.) Also show that if the output is linear in the input and x_0 is the initial vector, then $C_k x_0 = 0$ for all k if (2.10) is considered, and $C(t)x_0 = 0$ for all $t \geq t_0$ if (2.11) is considered.

2.6 Let $|A|_p$ be the l^p norm of the matrix $A=[a_{ij}(t)]$, that is, $|A|_p = |A(t)|_p = (\Sigma_{i,j}|a_{ij}(t)|^p)^{1/p}$. Under the hypothesis

$$\int_J |A(t)|_p^p\, dt < \infty \ ,$$

where $p>1$, prove that the infinite series (2.6) converges uniformly to $\Phi(t,t_0)$ on every bounded subinterval of J.
(*Hint*: Use Hölder inequality:

$$\int_J |A(t)B(t)|_1\, dt \leq (\int_J |A(t)|_p^p)^{1/p}(\int_J |B(t)|_q^q)^{1/q} \ ,$$

where $1/p+1/q=1$ and $1<p<\infty$.)

2.7 Discretize the continuous-time input-state equation

$$\begin{bmatrix} \dot{x}_1 \\ \dot{x}_2 \end{bmatrix} = \begin{bmatrix} 1 & -5 \\ 0 & -10 \end{bmatrix}\begin{bmatrix} x_1 \\ x_2 \end{bmatrix} + \begin{bmatrix} 1 \\ -1 \end{bmatrix}(t+1)$$

by using both methods discussed in Sect. 2.4 and compare both transition state equations. Try to bring $\begin{bmatrix} a \\ b \end{bmatrix}$ to the origin in both cases. Use $h=1/5$ and $1/10$.

2.8 Let $|A|_p$ be defined as in Exercise 2.6. Show that if A and B are matrices of the same dimension, then $|A+B|_p \leq |A|_p+|B|_p$ (called the triangle inequality).
(*Hint*: Use the Hölder inequality: For real numbers a_{ij} and b_{ij},

$$\sum_{i,j} |a_{ij}|\,|b_{ij}| \leq \left(\sum_{i,j} |a_{ij}|^p\right)^{1/p}\left(\sum_{i,j} |b_{ij}|^q\right)^{1/q}$$

where $1/p+1/q=1$ and $1<p<\infty$).

3. Controllability

The notion of controllability is introduced in this chapter. Both continuous- and discrete-time systems will be studied. If the system is time-invariant, then its controllability is completely determined by a constant matrix.

3.1 Control and Observation Equations

A linear system with continuous-time state-space description

$$\dot{x} = A(t)x + B(t)u$$
$$v = C(t)x + D(t)u \tag{3.1}$$

can be considered as a "control-observation" process, with $u = u(t)$ denoting the *control function* and $v = v(t)$ the observation function. Under the influence of the control u, the state vector $x = x(t)$ travels in the n-space $\mathbb{R}^n$ and traces a path in $\mathbb{R}^n$ as time increases in the allowable time interval. In order to give a more complete discussion, we always assume that the time interval J extends to positive infinity, and to apply the theory developed in Chap. 2, we also assume that the $n \times n$ system matrix $A = A(t)$ has continuous entries on J. If the admissible class of control functions u contains only piecewise continuous (or more generally bounded measurable) functions on J, then the entries of the control matrix $B = B(t)$ are allowed to be piecewise continuous (or more generally bounded measurable) functions; but if delta distributions are used as control "functions", then we must restrict the entries of the control matrix to continuous functions on J. The first equation in (3.1), namely the input-state relation, describes the control process and hence will be called the *control differential equation*. From Sect. 2.1, we know that this equation has an equivalent formulation

$$x(t) = \Phi(t, t_0)x(t_0) + \int_{t_0}^{t} \Phi(t, s)B(s)u(s)ds \tag{3.2}$$

which describes the path of travel of the state vector x under the influence of the control function u as the time parameter t increases starting at the initial time t_0. Since the transition matrix $\Phi(t, t_0)$ in the state-transition equation (3.2) is always nonsingular, the transition process is reversible; that is, multiplying both sides of

(3.2) by $\Phi^{-1}(t, t_0) = \Phi(t_0, t)$, we obtain the same equation with t and t_0 interchanged (although $t > t_0$). The second equation in the state-space description (3.1) will be called the *observation equation* since it describes the observation process. Of course analogous terminology and discussion apply to the discrete-time state-space description, but since the transition matrix in the discrete (or digital) model may turn out to be singular, a reversed transition may be impossible. We will postpone discussing the control properties of this model to the end of this chapter.

3.2 Controllability of Continuous-Time Linear Systems

The notion of controllability and complete controllability is introduced in this section. We first discuss controllability of a continuous-time linear system; the discrete-time setting being delayed to Sect. 3.4.

Definition 3.1 A linear system $\mathscr{S}$ with a state-space description given by (3.1) is said to be *controllable* if, starting from any position x_0 in $\mathbb{R}^n$, the state vector x at any initial time $t_0 \in J_0$ can be brought to the origin 0 in $\mathbb{R}^n$ in a finite amount of time by a certain control function u. In other words, the system $\mathscr{S}$ is *controllable* if for arbitrarily given $x_0 \in \mathbb{R}^n$ and $t_0 \in J$, there exists a $t_1 \geq t_0$ such that the integral equation

$$\Phi(t_1, t_0)x_0 + \int_{t_0}^{t_1} \Phi(t_1, s)B(s)u(s)ds = 0 \tag{3.3}$$

has a solution u in the admissible class of control functions.

Hence, to verify controllability, one has to prove the existence of both $t_1 \geq t_0$ and a control function u for any position x_0 in $\mathbb{R}^n$. Our first goal is to eliminate the difficulty imposed by the dependence of time on space by proving the existence of a "universal" finite time-interval. To do this we introduce the following subspaces. Let $t_0 \in J$ be fixed, and for each $t_1 \geq t_0$, let V_{t_1} be the collection of all x_0 in $\mathbb{R}^n$ such that (3.3) has an admissible solution u, and

$$V = \cup \{V_{t_1} : t_1 \geq t_0\} .$$

Then the above definition of controllability has the following equivalent statement.

Lemma 3.1 $\mathscr{S}$ *is controllable if and only if* $V = \mathbb{R}^n$.

It is clear that V and V_t, $t \geq t_0$, are all subspaces of $\mathbb{R}^n$ and that if (3.3) has a solution u and $t_2 \geq t_1$, then (3.3) with t_1 replaced by t_2 also has a solution (Exercise 3.1). Hence V_s is a subspace of V_t if $t \geq s \geq t_0$. Let $f(t)$ denote the dimension of V_t. Then f is a nondecreasing integer-valued function with

$$\lim_{t \to \infty} f(t) = \dim V \leq n .$$

By using the definition of limit, there is a $t^* \geq t_0$ such that $|f(t) - \dim V| < 1/2$ for all $t \geq t^*$, which implies immediately that $f(t^*) = \dim V$ and $V_{t^*} = V$. That is, we have proved the following result.

Theorem 3.1 *Let $\mathscr{S}$ be a linear system with the state-space description (3.1) and $t_0 \in J$. Then there exists a (finite) $t^* \geq t_0$ with $V_{t^*} = V$. Furthermore, the system $\mathscr{S}$ is controllable if and only if for any $x_0 \in \mathbb{R}^n$ the equation*

$$\Phi(t^*, t_0)x_0 + \int_{t_0}^{t^*} \Phi(t^*, s)B(s)u(s)ds = 0$$

has an admissible solution u.

The interval (t_0, t^*) will be called a *universal time-interval* for the system $\mathscr{S}$ with initial time t_0. As discussed earlier (Exercise 3.1), if (3.3) has a solution u with $t_1 < t^*$, then it has a solution when t_1 is replaced by t^*.

In the study of controllability, two linear transformations are of particular importance. They are

$$L_t u = \int_{t_0}^{t} \Phi(t, s)B(s)u(s)ds \quad \text{and} \tag{3.4}$$

$$Q_t = \int_{t_0}^{t} \Phi(t, s)B(s)B^T(s)\Phi^T(t, s)ds \ . \tag{3.5}$$

The first one maps the space of admissible control functions into $\mathbb{R}^n$ and the second one is an $n \times n$ matrix. We will next show that they have the same image. Using notation from linear algebra, we let "Im" denote "the image of" and "v" denote "the null space of".

Lemma 3.2 Im$\{L_t\}$ = Im$\{Q_t\}$ *for all $t \geq t_0$.*

We first show the easy direction. Let x be in Im$\{Q_t\}$. Then there is a $y \in \mathbb{R}^n$ such that

$$x = Q_t y = \int_{t_0}^{t} \Phi(t, s)B(s)u(s)ds = L_t u$$

with u defined by

$$u(s) = B^T(s)\Phi^T(t, s)y \ .$$

To establish the other direction, we first note that Q_t is symmetric so that Im$\{Q_t\}$ is orthogonal to vQ_t (Exercise 3.2). Hence, if x is not in Im$\{Q_t\}$, we can decompose x into $x = x_1 + x_2$ where $x_1 \in \text{Im}\{Q_t\}$ and $0 \neq x_2 \in vQ_t$, so that $x^T x_2 = x_1^T x_2 + x_2^T x_2 = x_2^T x_2 \neq 0$. If, on the other hand, x is in Im$\{L_t\}$, then there is some control function u with $L_t u = x$, so that

$$\int_{t_0}^{t} x_2^T \Phi(t, s)B(s)u(s)ds = x_2^T L_t u = x_2^T x \neq 0$$

and $x_2^T \Phi(t, s) B(s)$ cannot be 0. This contradicts the fact that

$$\int_{t_0}^{t} [x_2^T \Phi(t, s) B(s)] [x_2^T \Phi(t, s) B(s)]^T ds = x_2^T \left[\int_{t_0}^{t} \Phi(t, s) B(s) B^T(s) \Phi^T(t, s) ds \right] x_2$$

$$= x_2^T Q_t x_2 = 0 .$$

That is, if $x \notin \operatorname{Im}\{Q_t\}$, then $x \notin \operatorname{Im}\{L_t\}$ either, establishing the other direction of the lemma.

We are now ready to state an important result of controllability.

Theorem 3.2 *Let $\mathscr{S}$ be a continuous-time linear system with a universal time interval $(t_0, t^*) \subset J$. Then $\mathscr{S}$ is controllable with initial time t_0 if and only if the matrix Q_{t^*} is nonsingular.*

This result follows from Lemma 3.2 by using $t = t^*$ and the fact that $\Phi(t^*, t_0)$ is nonsingular (Exercise 3.3). It should be pointed out that in general it is impossible to determine the rank of the matrix Q_{t^*} since it is very difficult to decide how large t^* has to be. However, if the system and control matrices A and B, respectively, are constant matrices, then Q_{t^*} is nonsingular if and only if Q_t is nonsingular for any $t > t_0$ (Exercises 3.4 and 5). As a consequence of Theorem 3.2, we can extend the idea of controllability to "complete controllability".

3.3 Complete Controllability of Continuous-Time Linear Systems

We next discuss the notion of complete controllability.

Definition 3.2   A system $\mathscr{S}$ with state-space description (3.1) is said to be *completely controllable* if, starting from any position x_0 in $\mathbb{R}^n$, the state vector x at any initial time $t_0 \in J$ can be brought to any other position x_1 in $\mathbb{R}^n$ in a finite amount of time by a certain control function u. In other words, $\mathscr{S}$ is *completely controllable*, if for arbitrarily given x_0 and x_1 in $\mathbb{R}^n$ and $t_0 \in J$, there exists a $t_1 \geq t_0$ such that the integral equation

$$\Phi(t_1, t_0) x_0 + \int_{t_0}^{t_1} \Phi(t_1, s) B(s) u(s) ds = x_1$$

has a solution u in the admissible class of control functions.

It is important to observe that, at least in continuous-time state-space descriptions, there is no difference between controllability and complete controllability. It will be seen later that this result does not apply to discrete-time linear systems in general.

Theorem 3.3 *Let $\mathscr{S}$ be a continuous-time linear system. Then $\mathscr{S}$ is completely controllable if and only if it is controllable. Furthermore, if $(t_0, t^*) \subset J$ is a universal* *time-interval and x_0, x_1 are arbitrarily given position vectors in $\mathbb{R}^n$, then the equation*

$$\Phi(t^*, t_0)x_0 + \int_{t_0}^{t^*}\Phi(t^*, s)B(s)u(s)ds = x_1 \tag{3.6}$$

has an admissible solution u.

In fact we can prove more. Let (t_0, t^*) be a universal time-interval. We introduce a *universal control function* $u = u^*$ that brings the state vector x from any position y_0 to any other position y_1 in $\mathbb{R}^n$ defined by

$$u^*(t) = B^T(t)\Phi^T(t^*, t)Q_{t^*}^{-1}(y_1 - \Phi(t^*, t_0)y_0) \ .$$

This is possible since Q_{t^*} is nonsingular if the system is controllable by using Theorem 3.2.

Next, we consider the special cases where the $n \times n$ system matrix A and the $n \times p$ control matrix B are constant matrices. Under this setting, we introduce an $n \times pn$ "compound" matrix

$$M_{AB} = [B \ AB \ \ldots \ A^{n-1}B] \tag{3.7}$$

and give a more useful criterion for (complete) controllability.

Theorem 3.4   *A time-invariant (continuous-time) linear system $\mathscr{S}$ is (completely) controllable if and only if the $n \times pn$ matrix M_{AB} has rank n.*

To prove this theorem, let us first assume that the rank of M_{AB} is less than n, so that its n rows are linearly dependent. Hence, there is a nonzero n-vector a with $a^T M_{AB} = [0 \ \ldots \ 0]$, or equivalently, $a^T B = a^T AB = \ldots = a^T A^{n-1}B = 0$. An easy application of the Cayley-Hamilton Theorem now gives $a^T A^k B = 0$ for $k = 0, 1, 2, \ldots$, so that $a^T \exp[(t^* - s)A]B = 0$ also (Exercise 3.7). Hence,

$$a^T\left(e^{(t^* - t_0)A}y_0 + \int_{t_0}^{t^*}e^{(t^* - s)A}Bu(s)\,ds - y_1\right) = a^T e^{(t^* - t_0)A}y_0 - a^T y_1 \ .$$

Hence, there does not exist any control function u that can bring the state vector from the position $y_0 = 0$ to those positions y_1 with $a^T y_1 \neq 0$. In particular, the position $y_1 = a \neq 0$ cannot be reached from 0. Hence, (complete) controllability implies that M_{AB} has rank n. Conversely, let us now assume that M_{AB} has rank n, and contrary to what we must prove, that $\mathscr{S}$ is not controllable. Let (t_0, t^*) be a universal time-interval. Then from Theorem 3.2 we see that Q_{t^*} is singular so that there exists some nonzero $x_0 \in \mathbb{R}^n$ with $Q_{t^*}x_0 = 0$. Hence, since $\Phi(t, s) = \exp[(t - s)A]$, we have

$$\int_{t_0}^{t^*}(x_0^T e^{(t^* - s)A}B)(x_0^T e^{(t^* - s)A}B)^T ds = x_0^T Q_{t^*}x_0 = 0$$

so that

$$x_0^T e^{(t^*-s)A} B = 0$$

for $t_0 \leq s \leq t^*$. Taking the first $(n-1)$ derivatives with respect to s and then setting $s = t^*$, we have

$$x_0^T A^k B = 0, \quad k = 0, \ldots, n-1 \, ,$$

so that $x_0^T M_{AB} = 0$. This gives a row dependence relationship of the matrix M_{AB} contradicting the hypothesis that M_{AB} has rank n.

In view of Theorem 3.4, the matrix M_{AB} in (3.7) is called the *controllability matrix* of the time-invariant system.

3.4 Controllability and Complete Controllability of Discrete-Time Linear Systems

We now turn to a linear system $\mathscr{S}$ with a discrete-time state-space description

$$\begin{aligned} x_{k+1} &= A_k x_k + B_k u_k \\ v_k &= C_k x_k + D_k u_k \end{aligned} \tag{3.8}$$

where the first equation is called the *control difference equation* and the second will be called the *observation equation* in the next chapter. The state-transition equation can be written, by a change of index in (2.8), as

$$x_k = \Phi_{kj} x_j + \sum_{i=j+1}^{k} \Phi_{ki} B_{i-1} u_{i-1} \tag{3.9}$$

where the transition matrix is

$$\Phi_{kj} = A_{k-1} \ldots A_j, \quad k > j \tag{3.10}$$

with $\Phi_{kk} = I$, the identity matrix. Analogous to the continuous-time state-space description, we define "controllability" and "complete controllability" as follows:

Definition 3.3   A system $\mathscr{S}$ with a state-space description given by (3.8) is said to be *controllable* if, starting from any position y_0 in $\mathbb{R}^n$, the state sequence $\{x_k\}$, with any initial time l, can be brought to the origin by a certain control sequence $\{u_k\}$ in a finite number of discrete time steps. It is said to be *completely controllable*, if it can be brought to any preassigned position y_1 in $\mathbb{R}^n$. That is, $\mathscr{S}$ is controllable if for any y_0 in $\mathbb{R}^n$ and integer l, there exist an integer N and a sequence $\{u_k\}$ such that

$$\Phi_{Nl} y_0 + \sum_{k=l+1}^{N} \Phi_{Nk} B_{k-1} u_{k-1} = 0 \tag{3.11}$$

and is completely controllable if for an additional preassigned y_1 in $\mathbb{R}^n$, N and $\{u_k\}$ exist such that

$$\Phi_{Nl}y_0 + \sum_{k=l+1}^{N} \Phi_{Nk}B_{k-1}u_{k-1} = y_1 \ . \tag{3.12}$$

Unlike the continuous-time system, there are controllable discrete-time linear systems which are not completely controllable. An example of such a system is one whose system matrices A_k are all upper triangular matrices with zero diagonal elements and whose $n \times p$ control matrices $B_k = [b_{ij}(k)]$, $p \leq n$, satisfy $b_{ij}(k) = 0$ for $i \geq j$. For this system even the zero control sequence brings the state from any position to the origin but no control sequence can bring the origin to the position $[0 \ldots 0 \ 1]^T$ (Exercise 3.8).

Any discrete-time linear system, controllable or not, has a controllable subspace V of position vectors $y \in \mathbb{R}^n$ that can be brought to the origin by certain control sequence in a finite number of steps. Let V_k be the subspace of $y \in \mathbb{R}^n$ that can be brought to 0 in $k - l + 1$ steps. Then if y can be brought to zero in j_1 steps and $j_1 < j_2$, it can certainly be brought to zero in j_2 steps, it then follows that V_j is a subspace of V_k for $j \leq k$. Let f_k be the dimension of V_k. Since V is the union of all V_k, $k \geq l$, $\{f_k\}$ converges to dim V. Therefore there exists an $l^* > l$ such that $V_{l^*} = V$. $\{l, \ldots, l^*\}$ will be called a *universal discrete time-interval* of the system. This gives the following result.

Theorem 3.5 *Let $\mathscr{S}$ be a discrete-time linear system and l any integer. Then there exists an integer $l^* > l$ such that $V_{l^*} = V$. Furthermore, $\mathscr{S}$ is controllable if and only if for any y_0 in $\mathbb{R}^n$ there exists $\{u_l, \ldots, u_{l^*-1}\}$ such that (3.11) is satisfied with $N = l^*$.*

Let $\{l, \ldots, l^*\}$ be a universal discrete time-interval of the system, and analogous to the continuous-time setting, consider the matrix

$$R_{l^*} = \sum_{i=l+1}^{l^*} \Phi_{l^*i}B_{i-1}B_{i-1}^T\Phi_{l^*i}^T \ . \tag{3.13}$$

If R_{l^*} is nonsingular, a universal control sequence can be constructed following the proof of Theorem 3.3 to show that the system is completely controllable. On the other hand, if the transition matrices are nonsingular, controllability implies that R_{l^*} is nonsingular (Exercise 3.10). Hence, we have the following result.

Theorem 3.6 *Let $\mathscr{S}$ be a discrete-time linear system with initial time $k = l$ and nonsingular system matrices $A_l, \ldots, A_{l^*-1}$ where $\{l, \ldots, l^*\}$ is a universal discrete time-interval. Then $\mathscr{S}$ is completely controllable if and only if it is controllable.*

It is important to note that although the system matrices, and consequently the transition matrices, could be singular, it is still possible for the matrix R_{l^*} to be nonsingular. In fact, regardless of the singularity of $A_l, \ldots, A_{l^*-1}$, the nonsingularity of R_{l^*} characterizes the complete controllability of the discrete-time system.

Theorem 3.7 *A discrete-time linear system is completely controllable if and only if the matrix R_{l^*} is nonsingular.*

One direction of this statement follows by constructing a *universal control sequence* with the help of $R_{l^*}^{-1}$ (Exercise 3.10). To prove the other direction, we imitate the proof of Lemma 3.2 by investigating the image of the linear operator S_{l^*} defined by

$$S_{l^*}\{u_k\} = \sum_{k=\bar{l}+1}^{l^*} \Phi_{l^*k} B_{k-1} u_{k-1} . \tag{3.14}$$

Clearly, if the system is completely controllable so that any position in $\mathbb{R}^n$ can be "reached" from 0, then the image of S_{l^*} is all of $\mathbb{R}^n$. Hence, if one could show that the image of R_{l^*} is the same as that of S_{l^*}, then R_{l^*} would be full rank or nonsingular. The reader is left to complete the details (Exercise 3.15).

We now consider time-invariant systems. Again the controllability matrix

$$M_{AB} = [B \ AB \ \dots \ A^{n-1} B]$$

plays an important role in characterizing complete controllability.

Theorem 3.8 *A time-invariant discrete-time linear system is completely controllable if and only if its controllability matrix has full rank.*

Since we only consider constant system and control matrices A and B, the state-transition equation (3.9) becomes:

$$x_k = A^{k-l} x_l + \sum_{i=\bar{l}+1}^{k} A^{k-i} B u_{i-1} ,$$

where again l is picked as the initial time. In view of the Cayley-Hamilton Theorem, it is natural to choose $l^* = n + l$, n being the dimension of the square matrix A. That is, the state-transition equation becomes

$$M_{AB} \begin{bmatrix} u_{n+l-1} \\ \vdots \\ u_l \end{bmatrix} = -x_n + A^n x_l . \tag{3.15}$$

Hence, if any "position" x_n in $\mathbb{R}^n$ can be "reached" from $x_l = 0$, the range of M_{AB} is all of $\mathbb{R}^n$ so that it has full rank. Conversely, if the row rank of M_{AB} is full, then the sequence $\{u_l, \dots, u_{n+l-1}\}$ can be obtained for arbitrary initial and final states x_l and x_n, respectively, by solving (3.15). This completes the proof of the theorem.

As a bonus of the above argument, we see that $\{l, \dots, n+l\}$ is a universal discrete time-interval. That is, if the state vector x_k at a position y_0 in $\mathbb{R}^n$ cannot be brought to the origin in n steps, it can never be brought to the origin by any control sequence u_k no matter how long it takes.

Exercises

3.1 Let V_t be the collection of all x_0 in $\mathbb{R}^n$ that can be brought to the origin in continuous-time by certain control functions with initial time t_0 and terminal time t, and V be the union of all V_t. Prove that V and V_t are subspaces of $\mathbb{R}^n$. Also show that V_s is a subspace of V_t if and only if $s \leq t$ by showing that if x_0 can be brought to 0 at terminal time s, it can be brought to 0 at terminal time t.

3.2 Let R be a symmetric $n \times n$ matrix and consider R as a linear transformation of $\mathbb{R}^n$ into itself. Show that each x in $\mathbb{R}^n$ can be decomposed into $x = x_1 + x_2$ where x_1 is in $\mathrm{Im}\{R\}$ and x_2 is in vR and that this decomposition is unique in the sense that if x is zero then both x_1 and x_2 are zero, by first proving that $\mathrm{Im}\{R\} = (vR)^{\perp}$.

3.3 By applying Lemma 3.2 with $t = t^*$, prove Theorem 3.2.

3.4 Let

$$A = \begin{bmatrix} 0 & 1 \\ 0 & 0 \end{bmatrix} \quad \text{and} \quad B = \begin{bmatrix} 1 \\ 1 \end{bmatrix}.$$

Find Q_t and determine if the linear system is controllable.

3.5 Let

$$A = \begin{bmatrix} 0 & 1 \\ 0 & 0 \end{bmatrix} \quad \text{and} \quad B = \begin{bmatrix} a \\ b \end{bmatrix}.$$

Determine all values of a and b for which the linear system is controllable. Verify the statement that if Q_t is nonsingular for some t, it is also nonsingular for any $t > t_0$.

3.6 Let Q_{t^*} be nonsingular where (t_0, t^*) is a universal time-interval. Show that the universal control function

$$u^*(t) = B^T(t) \Phi^T(t^*, t) Q_{t^*}^{-1} [y_1 - \Phi(t^*, t_0) y_0]$$

brings x from y_0 to y_1. (This proves Theorem 3.3).

3.7 Let A be an $n \times n$ matrix. Show that if $a^T A^k = 0$ for $k = 0, \ldots, n-1$, then $a^T \exp(bA) = 0$ for any real number b and $a \in \mathbb{R}^n$.

3.8 Let $A_k = [a_{ij}(k)]$ be $n \times n$ and $B_k = [b_{ij}(k)]$ be $n \times p$ matrices where $p \leq n$ such that $a_{ij}(k) = b_{ij}(k) = 0$ if $i \geq j$. Show that the corresponding discrete-time linear system is controllable but not completely controllable. Also, verify that the system

$$x_{k+1} = \begin{bmatrix} 10 & 0 \\ -1 & 0 \end{bmatrix} x_k + \begin{bmatrix} -1 & 0 \\ 0.1 & 0 \end{bmatrix} u_k$$

$$x_0 = \begin{bmatrix} a \\ b \end{bmatrix}$$

is controllable but not completely controllable for any real numbers a and b.

3.9 Let

$$A_k = \begin{bmatrix} 0 & k \\ 0 & 0 \end{bmatrix} \quad \text{and} \quad B_k = B = \begin{bmatrix} 0 \\ 1 \end{bmatrix}.$$

Although the system matrices A_k are singular, show that the corresponding linear system is completely controllable and that any universal discrete time-interval is of "length" two.

3.10 Prove that if R_{l^*} is nonsingular then the corresponding linear system is controllable. Also show that if the state vector x_k can be brought from x_0 to the origin then $y_0 = -\Phi_{l^*l}x_0$ is in the image of R_{l^*}. This last statement shows that R_{l^*} is nonsingular since y_0 represents an arbitrary vector in $\mathbb{R}^n$.

3.11 By imitating the proof of Theorem 3.3 in Exercise 3.6, give a proof of Theorem 3.6.

3.12 Show that a universal discrete time-interval for a time-invariant system can be chosen such that its "length" does not exceed the order of the system matrix A. Give an example to show that this "length" cannot be shortened in general.

3.13 Let $\mathscr{S}$ be a linear system with the input-output relation $v'' + av' + bv = cu' + du$. Determine all values of a, b, c, and d for which this system is (completely) controllable.

3.14 Let $\mathscr{S}$ be a discrete linear system with the input-output relation $v_{k+2} + av_{k+1} + bv_k = u_{k+1} + cu_k$. Determine all values of a, b and c for which this system is controllable, and those values for which it is completely controllable.

3.15 Complete the proof of Theorem 3.7 by showing that R_{l^*} and S_{l^*} have the same image.

4. Observability and Dual Systems

In studying controllability or complete controllability of a linear system $\mathscr{S}$, only the control differential (or difference) equation in the state-space description of $\mathscr{S}$ has to be investigated. In this chapter the concept of "observability" is introduced and discussed. The problem is to deduce information of the initial state from knowledge of an input-output pair over a certain period of time. The importance of determining the initial state is that the state vector at any instant is also determined by using the state-transition equation. Since the output function is used in this process, the observation equation must also play an important role in the discussion.

4.1 Observability of Continuous-Time Linear Systems

Again we first consider the continuous-time model under the same basic assumptions on the time-interval J and the $n \times n$ and $n \times p$ matrices $A(t)$ and $B(t)$, respectively, as in the previous chapter. In addition, we require the entries of the $q \times n$ and $q \times p$ matrices $C(t)$ and $D(t)$, respectively, to be piecewise continuous (or more generally bounded measurable) functions on J.

We will say that a linear system $\mathscr{S}$ with the state-space description

$$\dot{x} = A(t)x + B(t)u$$

$$v = C(t)x + D(t)u \tag{4.1}$$

has the *observability property on an interval* $(t_0, t_1) \subset J$, if any input-output pair $(u(t), v(t))$, $t_0 \le t \le t_1$, uniquely determines an initial state $x(t_0)$.

Definition 4.1 A linear system $\mathscr{S}$ described by (4.1) is said to be *observable at an initial time* t_0 if it has the observability property on *some* interval (t_0, t_1) where $t_1 > t_0$. It is said to be *completely observable* or simply *observable* if it is observable at every initial time $t_0 \in J$.

Definition 4.2 A linear system $\mathscr{S}$ described by (4.1) is said to be *totally observable at an initial time* t_0 if it has the observability property on *every* interval (t_0, t_1) where $t_1 > t_0$. It is said to be *totally observable* if it is totally observable at every initial time $t_0 \in J$.

It is clear that every totally observable linear system is observable. But there are observable linear systems that are not totally observable. One example is a time-varying linear system with system and observation matrices given by

$$A = \begin{bmatrix} 0 & -1 \\ 0 & 0 \end{bmatrix} \quad \text{and} \quad C(t) = [1 \;\; 1 - |t-1|] \;,$$

respectively. This system is observable at every initial time $t_0 \geq 0$, totally observable at $t_0 \geq 1$, but not at any initial time between 0 and 1 (Exercise 4.1). Another interesting example is a linear system with the same system matrix A and with the observation matrix given by $[1 \;\; 1 + |t-1|]$. It can be shown that this system is totally observable at any initial time t_0 with $0 \leq t_0 < 1$ but is not observable at any $t_0 \geq 1$ (Exercise 4.2). To understand the observability of the above two linear systems and other time-varying systems in general, it is important to give an observability criterion. The matrix

$$P_t = \int_{t_0}^{t} \Phi^T(\tau, t_0) C^T(\tau) C(\tau) \Phi(\tau, t_0) \, d\tau \tag{4.2}$$

plays an important role for this purpose.

Theorem 4.1 *A linear system $\mathcal{S}$ described by (4.1) is observable at an initial time t_0 if and only if the square matrix P_t given by (4.2) is nonsingular for some value of $t > t_0$. In fact, it has the observability property on (t_0, t_1) if and only if P_{t_1} is nonsingular.*

Suppose that $\mathcal{S}$ is observable at t_0, and the zero input is used with output $v_0(t)$. Then there is a $t_1 > t_0$ such that the pair $(0, v_0(t))$, for $t_0 \leq t \leq t_1$, uniquely determines the initial state $x(t_0)$. Assume, contrary to what has to be proved, that P_t is singular for all $t > t_0$. Then, there is a nonzero x_0 (depending on t_1) such that

$$x_0^T P_{t_1} x_0 = 0 \;.$$

It therefore follows from (4.2) that

$$C(t)\Phi(t, t_0) x_0 = 0$$

for $t_0 \leq t \leq t_1$. However, from the state-transition equation with $u = 0$, we also have

$$v_0(t) = C(t)\Phi(t, t_0) x(t_0) \;,$$

so that $v_0(t) = C(t)\Phi(t, t_0)(x(t_0) + \alpha x_0)$ for any constant α, contradicting the fact that the pair $(0, v_0(t))$, $t_0 \leq t \leq t_1$, uniquely determines $x(t_0)$. To prove the converse, assume that P_{t_1} is nonsingular for some $t_1 > t_0$. Again from the state-transition equation, together with the control equation in (4.1), we have

$$C(t)\Phi(t, t_0)x(t_0) = v(t) - D(t)u(t) + \int_{t_0}^{t} C(t)\Phi(t, \tau)B(\tau)u(\tau)d\tau \;. \tag{4.3}$$

Multiplying both sides to the left by $\Phi^T(t, t_0)C^T(t)$ and integrating from t_0 to t_1, we have

$$P_{t_1}x(t_0) = \int_{t_0}^{t_1} \Phi^T(t, t_0)C^T(t)v(t)dt$$

$$- \int_{t_0}^{t_1} \Phi^T(t, t_0)C^T(t)D(t)u(t)dt$$

$$- \int_{t_0}^{t_1}\int_{t_0}^{t} \Phi^T(t, t_0)C^T(t)C(t)\Phi(t, \tau)B(\tau)u(\tau)d\tau dt \ .$$

Since P_{t_1} is nonsingular, $x(t_0)$ is uniquely determined by u and v over the time duration (t_0, t_1). This completes the proof of the theorem.

For time-invariant systems, we have a more useful observability criterion. Let A and C be constant $n \times n$ and $q \times n$ matrices and consider the $qn \times n$ compound matrix

$$N_{CA} = \begin{bmatrix} C \\ CA \\ \vdots \\ CA^{n-1} \end{bmatrix} . \tag{4.4}$$

In view of the following theorem, N_{CA} will be called the *observability matrix* of the linear system.

Theorem 4.2 *A time-invariant (continuous-time) linear system $\mathscr{S}$ is observable if and only if the $qn \times n$ matrix N_{CA} has rank n. Furthermore, if $\mathscr{S}$ is observable, it is also totally observable.*

Let us first assume that the rank of N_{CA} is less than n, so that the columns of N_{CA} are linearly dependent. That is, a nonzero n-vector a exists such that $N_{CA}a = 0$, or equivalently,

$$Ca = CAa = \ldots = CA^{n-1}a = 0 \ .$$

An application of the Cayley-Hamilton Theorem immediately gives $C\exp[(\tau - t_0)A] \, a = 0$ for all $\tau > t_0$ (Exercise 3.7). Now, multiplying to the left by the transpose of $C\exp[(\tau - t_0)A]$ and integrating from t_0 to t, we obtain

$$P_t a = 0$$

by using (4.2) and the fact that $\Phi(\tau, t_0) = \exp[(\tau - t_0)A]$. This holds for all $t > t_0$. That is, P_t is singular for all $t > t_0$ where t_0 was arbitrarily chosen from J. It follows from Theorem 4.1 that $\mathscr{S}$ is not observable at any initial time t_0 in J. Conversely, let us now assume that N_{CA} has rank n and let t_0 be arbitrarily chosen from J. We wish to show that $\mathcal{S}$ is not only observable at t_0, but is also totally observable there. That is, choosing any $t_1 > t_0$ and any input-output pair (u, v); we have to show that the initial state $x(t_0)$ is uniquely determined by $u(t)$ and $v(t)$ for $t_0 \le t \le t_1$. Let $\hat{x}(t_0)$ be any other initial state determined by $u(t)$ and $v(t)$ for $t_0 \le t \le t_1$. We must show that $\hat{x}(t_0) = x(t_0)$. Now since both $x(t_0)$ and $\hat{x}(t_0)$ satisfy (4.3) for $t_0 \le t \le t_1$, taking the difference of these two equations yields

$$C(t)\Phi(t, t_0)[x(t_0) - \hat{x}(t_0)] = Ce^{(t-t_0)A}[x(t_0) - \hat{x}(t_0)] = 0 \ ,$$

for $t_0 \le t \le t_1$. By taking the first $(n-1)$ derivatives with respect to t and setting $t = t_0$, we have

$$CA^k(x(t_0) - \hat{x}(t_0)) = 0, \quad k = 0, \ldots, n-1 \ ,$$

which is equivalent to $N_{CA}[x(t_0) - \hat{x}(t_0)] = 0$. Since N_{CA} has full column rank, we can conclude that $x(t_0)$ and $\hat{x}(t_0)$ are identical. This completes the proof of the theorem.

It is perhaps not very surprising that there is no distinction between observable and totally observable continuous-time time-invariant linear systems. It is important to point out, however, that for both time-varying and time-invariant discrete-time linear systems, total observability is in general much stronger than (complete) observability.

4.2 Observability of Discrete-Time Linear Systems

We now consider discrete-time linear systems. Let $\mathcal{S}$ be a discrete-time linear system with the state-space description

$$\begin{aligned}
x_{k+1} &= A_k x_k + B_k u_k \\
v_k &= C_k x_k + D_k u_k \ .
\end{aligned} \tag{4.5}$$

Analogous to the continuous-time case, $\mathcal{S}$ is said to have the *observability property* on a discrete time-interval $\{l, \ldots, m\}$, if any pair of input-output sequences (u_k, v_k), $k = l, \ldots, m$, uniquely determine an initial state x_l; or equivalently,

$$C_k \Phi_{kl} x_l = 0 \ , \tag{4.6}$$

$k = l, \ldots, m$, if and only if $x_l = 0$, where $\Phi_{ll} = I$ and $\Phi_{kl} = A_{k-1} \ldots A_l$ for $k > l$ (Exercise 4.5). Hence, it is clear that if $\mathcal{S}$ has the observability property on $\{l, \ldots, m\}$ it has the observability property on $\{l, \ldots, r\}$ for any $r \ge m$. For this reason the definitions for observability and total observability analogous to those in the continuous-time setting can be slightly modified.

Definition 4.3   A linear system $\mathscr{S}$ with a discrete-time state-space description (4.5) is said to be *observable at an initial time* l if there exists an $m > l$ such that whenever (4.6) is satisfied for $k = l, \ldots, m$ we must have $x_l = 0$. It is said to be *completely observable* or simply *observable* if it is observable at every initial time l.

Definition 4.4   A linear system $\mathscr{S}$ described by (4.5) is said to be *totally observable at an initial time* l, if whenever (4.6) is satisfied for $k = l$ and $l + 1$, we must have $x_l = 0$. It is said to be *totally observable* if it is totally observable at every initial time l.

To imitate the continuous setting, we again introduce an analogous matrix

$$L_m = \sum_{k=l+1}^{m} \Phi_{kl}^T C_k^T C_k \Phi_{kl} \tag{4.7}$$

and obtain an observability criterion.

Theorem 4.3   *A linear system $\mathscr{S}$ with a discrete-time state-space description given by* (4.5) *is observable at an initial time* l *if and only if there is an* $m > l$ *such that* L_m *is nonsingular.*

Since the proof of this theorem is similar to that of Theorem 4.1, we leave it as an exercise for the reader (Exercise 4.6). For time-invariant linear systems where $A_k = A$ and $C_k = C$ are $n \times n$ and $q \times n$ matrices, respectively, we have a more useful observability criterion.

Theorem 4.4   *A time-invariant (discrete-time) linear system $\mathscr{S}$ is observable if and only if the observability matrix N_{CA} defined by* (4.4) *has rank* n.

We again let the reader supply a proof for this result (Exercise 4.7). Since total observability is defined by two time-steps, we expect it to be characterized differently. This is shown in the following theorem.

Theorem 4.5   *A time-invariant (discrete-time) linear system $\mathscr{S}$ is totally observable if and only if the $2q \times n$ matrix*

$$T_{CA} = \begin{bmatrix} C \\ CA \end{bmatrix}$$

has rank n.

We call T_{CA} the *total observability matrix* of the discrete-time system. As a consequence of this theorem, we note that a discrete-time linear system that has the number of rows in its observation matrix less than half of the order of its system matrix is never totally observable. The proof of the above theorem follows from the definition of total observability (Exercise 4.8).

For example, if the system and observation matrices are, respectively,

$$A = \begin{bmatrix} 0 & 0 & 0 \\ 1 & 0 & 0 \\ 0 & 1 & 1 \end{bmatrix}, \qquad C = [0 \ \ 0 \ \ 1] \qquad \text{then}$$

$$N_{CA} = \begin{bmatrix} 0 & 0 & 1 \\ 0 & 1 & 1 \\ 1 & 1 & 1 \end{bmatrix}, \quad \text{and} \quad T_{CA} = \begin{bmatrix} 0 & 0 & 1 \\ 0 & 1 & 1 \end{bmatrix}$$

have ranks 3 and 2 respectively, so that the corresponding discrete-time linear system is completely but not totally observable.

4.3 Duality of Linear Systems

An interesting resemblance between a completely controllable *time-invariant* linear system and a completely observable one (either continuous- or discrete-time) is that they have very similar characterizations in terms of the controllability matrix M_{AB} and the observability matrix N_{CA}, respectively. In fact, the two continuous-time linear systems

$$\mathcal{S}_{\mathrm{c}}: \begin{cases} \dot{x} = Ax + Bu \\ v = Cx + Du \end{cases} \qquad \text{and}$$

$$\tilde{\mathcal{S}}_{\mathrm{c}}: \begin{cases} \dot{x} = A^T x + C^T \tilde{u} \\ \tilde{v} = B^T x + \tilde{D} \tilde{u} \ , \end{cases}$$

where A, B, and C are constant matrices, are "dual" to each other in the sense that the controllability matrix of $\mathcal{S}_{\mathrm{c}}$ is the transpose of the observability matrix of $\tilde{\mathcal{S}}_{\mathrm{c}}$, and the observability matrix of $\mathcal{S}_{\mathrm{c}}$ is the transpose of the controllability matrix of $\tilde{\mathcal{S}}_{\mathrm{c}}$. The same duality statement holds for the two discrete-time linear systems

$$\mathcal{S}_{\mathrm{d}}: \begin{cases} x_{k+1} = Ax_k + Bu_k \\ v_k = Cx_k + Du_k \end{cases} \qquad \text{and}$$

$$\tilde{\mathcal{S}}_{\mathrm{d}}: \begin{cases} x_{k+1} = A^T x_k + C^T \tilde{u}_k \\ \tilde{v}_k = B^T x_k + \tilde{D} \tilde{u}_k \ . \end{cases}$$

Hence, we obtain the following duality phenomenon by an immediate application of Theorems 3.4 and 4.2.

Theorem 4.6 *The two continuous-time linear systems $\mathscr{S}_c$ and $\mathscr{\tilde{S}}_c$ described above are dual to each other in the sense that $\mathscr{S}_c$ is completely controllable if and only if $\mathscr{\tilde{S}}_c$ is completely observable, and $\mathscr{S}_c$ is completely observable if and only if $\mathscr{\tilde{S}}_c$ is completely controllable. The same statement holds for the pair of discrete-time linear systems $\mathscr{S}_d$ and $\mathscr{\tilde{S}}_d$.*

The formulation of a "dual system" for the time-varying setting is more complicated. We first need the following result.

Lemma 4.1 *Let $\Phi(t, s)$ and $\Psi(t, s)$ be the transition matrices of $A(t)$ and $-A^T(t)$ respectively. Then $\Psi^T(s, t) = \Phi(t, s)$.*

To prove this result, we first differentiate the identity $\Psi(t, s)\,\Psi(s, t) = I$ with respect to t and obtain

$$\Psi_1(t, s)\,\Psi(s, t) + \Psi(t, s)\Psi_2(s, t) = 0 ,$$

where the subscripts 1 and 2 indicate the partial derivatives with respect to the first and second variables. Hence,

$$-A^T(t)\Psi(t, s)\Psi(s, t) + \Psi(t, s)\Psi_2(s, t) = 0, \quad \text{or}$$

$$\Psi_2(s, t) = \Psi(s, t)A^T(t)$$

and the lemma follows by taking the transpose of both sides of this identity.

We are now ready to formulate the dual time-varying systems. Let

$$\mathscr{S}_c : \begin{cases} \dot{x} = A(t)x + B(t)u \\ v = C(t)x + D(t)u \end{cases} \quad \text{and}$$

$$\mathscr{\tilde{S}}_c : \begin{cases} \dot{x} = -A^T(t)x + C^T(t)\tilde{u} \\ \tilde{v} = B^T(t)x + \tilde{D}(t)\tilde{u} . \end{cases}$$

Then we have the following duality result.

Theorem 4.7 *$\mathscr{S}_c$ is controllable with a universal time-interval (t_0, t^*), where $t^* > t_0$, if and only if $\mathscr{\tilde{S}}_c$ has the observability property on (t_0, t^*). Also, $\mathscr{S}_c$ has the observability property on (t_0, t_1), where $t_1 > t_0$, if and only if $\mathscr{\tilde{S}}_c$ is controllable with (t_0, t_1) as a universal time-interval.*

The proof of this result follows from Theorems 3.2 and 4.1 by applying Lemma 4.1 and relating the matrix

$$\Phi(t_0, t^*)Q_{t^*}\Phi^T(t_0, t^*)$$

to the P_{t^*} matrix

$$\int_{t_0}^{t^*} \Psi^T(t, t_0) B(t) B^T(t) \Psi(t, t_0)\, dt$$

of the system $\mathcal{S}_c$ (Exercise 4.9).

The negative sign in front of $A^T(t)$ in the state-space description of $\mathcal{S}_c$ does not cause inconsistency in the event that A, B, and C are constant matrices. The reason is that the matrices

$$\tilde{M}_{AB} = [B \quad -AB \quad \ldots \quad (-1)^{n-1} A^{n-1} B], \quad \tilde{N}_{CA} = \begin{bmatrix} C \\ -CA \\ \vdots \\ (-1)^{n-1} CA^{n-1} \end{bmatrix}$$

have the same ranks as M_{AB} and N_{CA}, respectively.

4.4   Dual Time-Varying Discrete-Time Linear Systems

For discrete-time linear systems, we do not need the negative sign in formulating the dual systems. We require, however, that the matrices A_k are nonsingular for $k = l, \ldots, l^* - 1$, instead (Theorems 3.6 and 7). Consider

$$\mathcal{S}_d : \begin{cases} x_{k+1} = A_k x_k + B_k u_k \\ v_k = C_k x_k + D_k u_k \end{cases} \quad \text{and}$$

$$\tilde{\mathcal{S}}_d : \begin{cases} x_{k+1} = (A_k^{-1})^T x_k + C_{k+1}^T \tilde{u}_k \\ \tilde{v}_k = B_{k-1}^T x_k + \tilde{D}_k \tilde{u}_k \ . \end{cases}$$

The following duality statement can be obtained by using the characterization matrices R_{l^*} and L_m (Exercise 4.10).

Theorem 4.8.   *Let $\mathcal{S}_d$ and $\tilde{\mathcal{S}}_d$ be the time-varying systems described above and suppose that $A_1, \ldots, A_{l^*-1}$ are nonsingular. Then $\mathcal{S}_d$ is completely controllable with a universal discrete time-interval $\{l, \ldots, l^*\}$ if and only if $\tilde{\mathcal{S}}_d$ has the observability property on $\{l, \ldots, l^*\}$. Also, $\mathcal{S}_d$ has the observability property on a discrete time-interval $\{l, \ldots, m\}$ if and only if $\tilde{\mathcal{S}}_d$ is completely controllable with $\{l, \ldots, m\}$ as a universal time-interval.*

We remark that in the special case where $A_l = \ldots = A_{l^*-1} = A$ is nonsingular, then Theorem 4.8 reduces to the last statement of Theorem 4.6 (Exercise 4.13).

Exercises

4.1 Let the system and observation matrices of a continuous-time linear system be

$$\begin{bmatrix} 0 & -1 \\ 0 & 0 \end{bmatrix} \quad \text{and} \quad [1 \quad 1-|t-1|] \; ,$$

respectively. Verify that this system is completely observable but not totally observable at any initial time less than 1.

4.2 In the above exercise, if the observation matrix is now changed to $[1 \quad 1+|t-1|]$, then verify that the new system is totally observable at any initial time t_0 where $0 \le t_0 < 1$ but is not even observable at any initial time $t_0 \ge 1$.

4.3 Find all values of a and b for which the linear systems with input-output relations given by $v'' - v' + v = au' + bu$ is observable.

4.4 Let

$$A = \begin{bmatrix} 0 & 0 \\ 1 & 0 \end{bmatrix} \quad \text{and} \quad C = [a \quad b].$$

Find P_t and N_{CA}. Compare the observability criteria in terms of these two matrices by showing that the same values of a and b are determined in each case.

4.5 Prove that the linear system described in (4.5) has the observability property on the discrete time-interval $\{l, \ldots, m\}$ if and only if $x_l = 0$ whenever (4.6) holds for $k = l, \ldots, m$.

4.6 Provide a proof for Theorem 4.3 by imitating that of Theorem 4.1.

4.7 Prove Theorem 4.4.

4.8 Prove that Theorem 4.5 is a direct consequence of the definition of total observability for discrete-time systems.

4.9 Supply the detail of the proof of Theorem 4.7.

4.10 Prove Theorem 4.8.

4.11 Let

$$A = \begin{bmatrix} 1 & 2 & 0 \\ 0 & 1 & 0 \\ -1 & 0 & a \end{bmatrix} \quad \text{and} \quad C = \begin{bmatrix} 1 & b & 1 \\ 0 & 0 & c \end{bmatrix}.$$

Determine all values of a, b and c for which the corresponding discrete-time linear system is completely observable and those values for which it is totally observable.

4.12 Consider a discrete-time linear system with input-output relations given by $v_{k+3} + av_{k+2} + bv_{k+1} + v_k = u_{k+1} + u_k$. Determine all values of a and b

for which the system is completely observable. Give the input-output relations for its dual system and determine all values of a and b for which the dual system is completely observable.

4.13 Let A be a nonsingular constant square matrix. Show that the two (continuous- or discrete-time) linear systems with the same constant observation matrix C and system matrices A and A^{-1}, respectively, are both observable if one of them is observable. The analogous statement holds for the controllability.

5. Time-Invariant Linear Systems

Time-invariant systems have many important properties which are useful in applications that time-varying systems do not possess. This chapter will be devoted to the study of some of their structural properties. In particular, the relationship between their state-space descriptions and transfer functions obtained by using Laplace or z-transforms will be discussed.

5.1 Preliminary Remarks

Before we concentrate on time-invariant systems, three items which are also valid for time-varying systems shoud be noted. These remarks will apply to both continuous- and discrete-time descriptions, although we only consider the continuous-time setting. The discrete-time analog is left as an exercise for the reader (Exercise 5.4).

Remark 5.1   The results on complete controllability and observability obtained in the previous two chapters seem to depend on the state-space descriptions of the linear systems; namely, on the matrices $A(t)$, $B(t)$, and $C(t)$. We note, however, that this dependence can be eliminated among the class of all state-space descriptions with the same cardinalities in state variables and input and output components, as long as the state vectors are nonsingular transformations of one another. More precisely, if G is any nonsingular constant matrix and the state vector x is changed to y by $y = G^{-1}x$, then the matrices $A(t)$, $B(t)$, and $C(t)$ are automatically changed to $\tilde{A}(t) = G^{-1}A(t)G$, $\tilde{B}(t) = G^{-1}B(t)$, and $\tilde{C}(t) = CG$, respectively. Hence, it is easy to see that if the transition matrix of the original state-space description is $\Phi(t,s)$, then the transition matrix of the transformed description can be written as $\tilde{\Phi}(t, s) = G^{-1}\Phi(t, s)G$, and it follows that the matrices $\tilde{Q}_{t^*}$ and $\tilde{P}_t$, which are used to give controllability and observability criteria for the transformed description as Q_{t^*} and P_t are for the original description, have the same ranks as Q_{t^*} and P_t, respectively, so that Theorems 3.2 and 4.1 tell us that controllability and observability properties are preserved (Exercise 5.1).

Remark 5.2   The transfer matrix $D(t)$ is certainly not useful in the study of controllability, and does not appear even in our discussion of observability. In fact, there is no loss of generality in assuming that $D(t)$ is zero and this we will do in this chapter (Exercise 5.2).

Remark 5.3 On the other hand the control equation in the state-space description can be slightly extended to include a vector-valued function, namely

$$\dot{x} = A(t)x + B(t)u + f(t) ,\qquad\qquad (5.1)$$

where $f(t)$ is a fixed $n \times 1$ matrix with piecewise continuous (or more generally bounded measurable) functions in all entries, without changing the controllability and observability properties (Exercise 5.3).

5.2 The Kalman Canonical Decomposition

We are now ready to study time-invariant linear systems. Let A, B, and C be constant $n \times n$, $n \times p$ and $q \times n$ matrices, respectively. These are of course the corresponding system, control, and observation matrices of the state-space descriptions of the linear system. Also, let the controllability and observability matrices be

$$M_{AB} = [B \quad AB \ldots A^{n-1}B] \quad \text{and}$$

$$N_{CA} = \begin{bmatrix} C \\ CA \\ \vdots \\ CA^{n-1} \end{bmatrix},$$

respectively. Recall that for both continuous- and discrete-time descriptions, these two matrices characterize complete controllability and observability in terms of the fullness of their ranks. Hence, if a system is not completely controllable or observable, it is natural to work with the matrices M_{AB} and N_{CA} to obtain a partition of some linear combination, which we will call "mixing", of the state variables into subsystems that have the appropriate complete controllability and observability properties. In addition, since only these two matrices will be considered, the following discussion will hold both for continuous- and discrete-time state-space descriptions.

Let sp M_{AB} denote the algebraic span of the column vectors of M_{AB} and sp N_{CA}^T that of the column vectors of N_{CA}^T. Next, let n_2 be the dimension of sp $M_{AB} \cap$ sp N_{CA}^T. It will be seen that n_2 is the number of state-variables, after

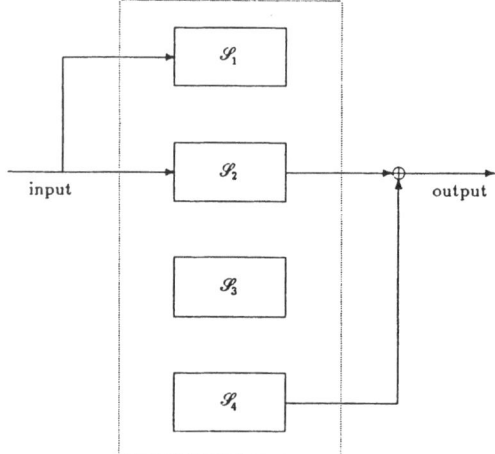

Fig. 5.1. Linear System $\mathscr{S}$

some "mixing", that constitute a largest subsystem $\mathscr{S}_2$ which is both completely controllable and observable. Also set

$$n_1 = \dim(\operatorname{sp} M_{AB}) - n_2 \; ,$$

$$n_4 = \dim(\operatorname{sp} N_{CA}^T) - n_2 \; , \quad \text{and}$$

$$n_3 = n - n_1 - n_2 - n_4 \; .$$

Clearly, $n_1, \ldots, n_4$ are all non-negative integers. It is believable that n_1 is the dimension of a subsystem $\mathscr{S}_1$ which is completely controllable but has zero output, and n_4 the number of state variables constituting a subsystem $\mathscr{S}_4$ which has zero control matrix but is observable. This is usually called the *Kalman Canonical Decomposition* (Fig. 5.1). However, to the best of our knowledge, there is no complete proof in the literature that $\mathscr{S}_1$ is completely controllable and $\mathscr{S}_4$ is observable. Further discussion on this topic is delayed to Chap. 10.

Let $\{e_1, \ldots, e_n\}$ be an orthonormal basis of $\mathbb{R}^n$ so constructed that $\{e_1, \ldots, e_{n_1+n_2}\}$ is a basis of $\operatorname{sp} M_{AB}$, $\{e_{n_1+1}, \ldots, e_{n_1+n_2}\}$ a basis of $\operatorname{sp} M_{AB} \cap \operatorname{sp} N_{CA}^T$, and $\{e_{n_1+1}, \ldots, e_{n_1+n_2}, e_{n_1+n_2+n_3+1}, \ldots, e_n\}$ a basis of $\operatorname{sp} N_{CA}^T$. We also consider the corresponding unitary matrix

$$U = [e_1 \ldots e_n]$$

whose jth column is e_j. This matrix can be considered as a nonsingular transformation that describes the first stage in "mixing" of the state variables briefly discussed above and in more detail later. This "mixing" procedure will put the transformed system matrix in the desired decomposable form. However, we will see later that this is not sufficient to ensure that the uncoupled subsystems have the desired controllability or observability properties. A second stage is required. Anyway, at present, the state vector $x(x_k$ for the corresponding discrete-time setting) is transformed to a state vector y (y_k for the discrete-time setting) defined by

$$y = U^{-1}x = U^T x \ .$$

Hence, the corresponding transformed system, control, and observation matrices are

$$\tilde{A} = U^T A U, \ \tilde{B} = U^T B, \quad \text{and} \quad \tilde{C} = CU \ ,$$

respectively.

We now collect some important consequences resulting from this transformation. Let us first recall a terminology from linear algebra: A subspace W of $\mathbb{R}^n$ is called an *invariant subspace* of $\mathbb{R}^n$ under a transformation L if Lx is in W for all x in W. In the following, we will identify certain invariant subspaces under the transformations A and A^T. For convenience, we denote the algebraic spans of $\{e_1, \ldots, e_{n_1}\}$ $\{e_{n_1+1}, \ldots, e_{n_1+n_2}\}$, $\{e_{n_1+n_2+1}, \ldots, e_{n_1+n_2+n_3}\}$, and $\{e_{n_1+n_2+n_3+1}, \ldots, e_n\}$ by V_1, V_2, V_3, and V_4 respectively. Hence, we have

$$\text{sp } M_{AB} = V_1 \oplus V_2, \quad \text{sp } M_{AB} \cap \text{sp } N_{CA}^T = V_2, \quad \text{sp } N_{CA}^T = V_2 \oplus V_4 \ .$$

Lemma 5.1 V_1 *and* $\text{sp } M_{AB}$ *are invariant subspaces of* $\mathbb{R}^n$ *under the transformation* A, *while* V_4 *and* $\text{sp } N_{CA}^T$ *are invariant subspaces under* A^T.

We only verify the first half and leave the second half as an exercise for the reader (Exercise 5.5). If x is in $\text{sp } M_{AB}$, then x is a linear combination of the columns of $B, AB, \ldots, A^{n-1}B$, so that Ax is a linear combination of the columns of $AB, \ldots, A^nB$. Hence, by the Cayley-Hamilton theorem, Ax is a linear combination of the columns of $B, AB, \ldots, A^{n-1}B$ again. That is, $\text{sp } M_{AB}$ is an invariant subspace of $\mathbb{R}^n$ under A. By the same argument, we see that $\text{sp } N_{CA}^T$ is an invariant subspace under A^T. Now let x be in V_1. Then x is in $\text{sp } M_{AB}$ so that Ax is also in $\text{sp } M_{AB} = V_1 \oplus V_2$. That is, $Ax = x_1 + x_2$ where x_1 is in V_1 and x_2 is in V_2. Since V_2 is a subspace of $\text{sp } N_{CA}^T$, $A^T x_2$ is also in $\text{sp } N_{CA}^T$. Hence, using the orthogonality between the vectors in V_1 and V_2, and the orthogonality between those in V_1 and $\text{sp } N_{CA}^T = V_2 \oplus V_4$ consecutively, we have

$$x_2^T x_2 = (x_1 + x_2)^T x_2$$

$$= (Ax)^T x_2 = x^T A^T x_2 = 0 \ .$$

That is, $x_2 = 0$, or $Ax = x_1$ which is in V_1. This shows that V_1 is an invariant subspace under A.

We next relate the images of e_j under A in terms of the basis $\{e_i\}$, using the coefficients from the entries of $\tilde{A}$. Write $\tilde{A} = [\tilde{a}_{ij}]$, $1 \le i, j \le n$. We have the following:

Lemma 5.2   *For each* $j = 1, \ldots, n,$

$$Ae_j = \tilde{a}_{1j}e_1 + \ldots + \tilde{a}_{nj}e_n . \tag{5.2}$$

The proof of this result is immediate from the identity $AU = U\tilde{A}$ $= [e_1 \ldots e_n]\tilde{A}$, since the vector on the left-hand side of (5.2) is the jth column of AU and the vector on the right-hand side of (5.2) is the jth column of $[e_1 \ldots e_n]A$.

We now return to the transformation $y = U^T x$ and show that the transformed state-space description has the desired decomposable form. Writing

$$y = \begin{bmatrix} y_1 \\ y_2 \\ y_3 \\ y_4 \end{bmatrix} \begin{matrix} \} n_1 \text{ components} \\ \} n_2 \text{ components} \\ \} n_3 \text{ components} \\ \} n_4 \text{ components} \end{matrix}$$

we can state the following decomposition result. Only the notation of a continuous-time system is used, and as usual, an extra subscript is required for the corresponding discrete-time system.

Theorem 5.1   *Every time-invariant linear system $\mathscr{S}$ whose transfer matrix D in its state-space description vanishes has a (nonsingular) unitary transformation $y = U^T x$ such that the transformed system, control, and observation matrices are of the form*

$$\tilde{A} = \begin{bmatrix} A_{11} & A_{12} & A_{13} & A_{14} \\ 0 & A_{22} & 0 & A_{24} \\ 0 & 0 & A_{33} & A_{34} \\ 0 & 0 & 0 & A_{44} \end{bmatrix} \begin{matrix} \} n_1 \\ \} n_2 \\ \} n_3 \\ \} n_4 \end{matrix} ,$$
$$\quad\;\; \underbrace{n_1} \;\; \underbrace{n_2} \;\; \underbrace{n_3} \;\; \underbrace{n_4}$$

$$\tilde{B} = \begin{bmatrix} B_1 \\ B_2 \\ 0 \\ 0 \end{bmatrix} \begin{matrix} \} n_1 \\ \} n_2 \\ \} n_3 \\ \} n_4 \end{matrix} \quad and \quad \tilde{C} = [\; \underset{n_1}{0} \quad \underset{n_2}{C_2} \quad \underset{n_3}{0} \quad \underset{n_4}{C_4} \;] .$$

Consequently, the transformed state-space description

$$\begin{cases} \dot{y} = \tilde{A}y + \tilde{B}u \\ v = \tilde{C}y \end{cases}$$

of $\mathscr{S}$ can be decomposed into four subsystems:

$$\mathscr{S}_1 : \begin{cases} \dot{y}_1 = A_{11}y_1 + B_1 u + f_1 \\ v = 0y_1 = 0 \end{cases}$$

with $f_1 = A_{12}y_2 + A_{13}y_3 + A_{14}y_4$,

$$\mathscr{S}_2 : \begin{cases} \dot{y}_2 = A_{22}y_2 + B_2 u + f_2 \\ v = c_2 y_2 \end{cases}$$

with $f_2 = A_{24}y_4$,

$$\mathscr{S}_3 : \begin{cases} \dot{y}_3 = A_{33}y_3 + 0u + f_3 = A_{33}y_3 + f_3 \\ v = 0y_3 = 0 \end{cases}$$

with $f_3 = A_{34}y_4$, and

$$\mathscr{S}_4 : \begin{cases} \dot{y}_4 = A_{44}y_4 + 0u = A_{44}y_4 \\ v = C_4 y_4 \end{cases}$$

where $\mathscr{S}_1$ has zero (or no) output for observation, $\mathscr{S}_2$ is both completely controllable and observable, $\mathscr{S}_3$ is not influenced by any control u and has no output, and $\mathscr{S}_4$ is not influenced by any control function.

It is important to note that although the combined ($\mathscr{S}_1$ and $\mathscr{S}_2$) system with system matrix

$$\begin{bmatrix} A_{11} & A_{12} \\ 0 & A_{22} \end{bmatrix}$$

is completely controllable, the subsystem $\mathscr{S}_1$ may not be controllable. This can be seen from the following example. Consider

$$A = \begin{bmatrix} 1 & 1 & 0 & 0 \\ 0 & 1 & 0 & 0 \\ 0 & 0 & 0 & 0 \\ 0 & 0 & 0 & 2 \end{bmatrix} \tag{5.3}$$

$$B = \begin{bmatrix} 0 \\ 1 \\ 0 \\ 0 \end{bmatrix} \quad \text{and} \quad C = [0 \ \ 1 \ \ 0 \ \ 1] \ .$$

As it stands, this is already in the desired decomposed form with $n_1 = n_2 = n_3 = n_4 = 1$. The subsystem $\mathscr{S}_2$ is clearly both completely controllable and observable, and the combined subsystem of $\mathscr{S}_1$ and $\mathscr{S}_2$ with

$$\begin{bmatrix} A_{11} & A_{12} \\ 0 & A_{22} \end{bmatrix} = \begin{bmatrix} 1 & 1 \\ 0 & 1 \end{bmatrix}$$

and control matrix $[0 \quad 1]^T$ is also completely controllable, since the controllability matrix is

$$\begin{bmatrix} 0 & 1 \\ 1 & 1 \end{bmatrix} \tag{5.4}$$

which is of full rank. However, the subsystem $\mathscr{S}_1$ is *not* controllable! Moreover, *no* unitary transformation *can* make $\mathscr{S}_1$ controllable (Exercise 5.6). Therefore, in general, a nonsingular (non-unitary) transformation is necessary. In this example, the transformation

$$G = \begin{bmatrix} 1 & 1 & 0 & 0 \\ 0 & 1 & 0 & 0 \\ 0 & 0 & 1 & 0 \\ 0 & 0 & 0 & 1 \end{bmatrix} \tag{5.5}$$

can do the job. We leave the detail as an exercise (Exercise 5.7).

We also point out that the dimensions $n_1, \ldots, n_4$ of the subsystems in the above theorem are independent of any nonsingular transformation (Exercise 5.8). For unitary transformations, this is clear. In fact, if W is any unitary $n \times n$ matrix and $\hat{A} = W^T A W$, $\hat{B} = W^T B$, and $\hat{C} = CW$, then the dimensions of the subspaces sp $M_{\hat{A}\hat{B}} \cap$ sp $N_{\hat{C}\hat{A}}^T$, sp $M_{\hat{A}\hat{B}}$, and sp $N_{\hat{C}\hat{A}}^T$ of $\mathbb{R}^n$ are clearly n_2, $n_1 + n_2$, and $n_4 + n_2$, respectively. In addition, we note that the vectors f_1, f_2, f_3 in the state-space descriptions of the subsystems do not change the controllability and observability properties as discussed in Remark 5.3, and the transfer matrix D does not play any role in this discussion (Remark 5.2). For convenience D was assumed to be the zero matrix in the above theorem. It is also worth mentioning that the nonsingular transformation U does not change the controllability and observability properties of the original state-space descriptions as observed in Remark 5.1.

To verify the structure of the matrix $\tilde{A}$ in the statement of Theorem 5.1, note that for $1 \leq j \leq n_1$, $Ae_j \in V_1$ by Lemma 5.1. Hence, comparing with the expression (5.2) in Lemma 5.2, we see that $\tilde{a}_{ij} = 0$ for $i = n_1 + 1, \ldots, n$ ($1 \leq j \leq n_1$). This shows that the first n_1 columns of $\tilde{A}$ have the block structure described in the theorem. To verify the structure of the second column block, we consider $n_1 + 1 \leq j \leq n_1 + n_2$ and note that Ae_j is in sp $M_{AB} = V_1 \oplus V_2$ from Lemma 5.1, so that again comparing with expression (5.2), we see that $\tilde{a}_{ij} = 0$ for $i = n_1 + n_2 + 1, \ldots, n$. For $n_1 + n_2 + 1 \leq j \leq n_1 + n_2 + n_3$, e_j is in V_3 and hence is orthogonal to any y in $V_2 \oplus V_4 = $ sp N_{CA}^T. But since sp N_{CA}^T is an invariant subspace of $\mathbb{R}^n$ under A^T, we see that $A^T y$ is also orthogonal to e_j, so that $(Ae_j)^T y = e_j^T A^T y = 0$, and Ae_j is in the orthogonal complement of sp N_{CA}^T. This shows that Ae_j is in $V_1 \oplus V_3$, which yields the zero structure of the third column block of $\tilde{A}$.

The zero structures of $\tilde{B}$ and $\tilde{C}$ again follow from orthogonality. Indeed, since the columns of B are in $V_1 \oplus V_2$, they are orthogonal to V_3 and V_4 so that the identity $\tilde{B} = U^T B = [e_1 \ldots e_n]^T B$ yields the described structure of $\tilde{B}$. Also, since the columns of C^T are in $V_2 \oplus V_4$ and $\tilde{C} = CU$, the first and third column blocks of $\tilde{C}$ must be zero. To verify the complete controllability and observability of the subsystem $\mathscr{S}_2$ in Theorem 5.1, one simply checks that the controllability and observability matrices are of full rank. In fact, it can also be shown that the combined $\mathscr{S}_1$ and $\mathscr{S}_2$ subsystem is completely controllable and the combined $\mathscr{S}_2$ and $\mathscr{S}_4$ subsystem is observable (Exercise 5.9).

5.3 Transfer Functions

Our next goal is to relate the study of state-space descriptions to that of the *transfer functions* which constitute the main tool in classical control theory. Recall that if $f(t)$ is a vector- (or matrix-) valued function defined on the time interval that extends from 0 to $+\infty$ such that each component (or entry) of $f(t)$ is a piecewise continuous (or more generally bounded measurable) function with at most exponential growth, then its Laplace transform is defined by

$$F(s) = (\mathscr{L}f)(s) = \int_0^\infty e^{-st} f(t)\, dt \ , \tag{5.6}$$

where, as usual, integration is performed entry-wise. This transformation takes $f(t)$ from the time domain to the frequency s-domain. The most important property for our purpose is that it changes an ordinary differential equation into an algebraic equation via

$$(\mathscr{L}f')(s) = sF(s) - f(0) \tag{5.7}$$

etc. Similarly, the z-transform maps a vector- (or matrix-) valued infinite sequence $\{g_k\}$, $k = 0, 1, \ldots$, to a (perhaps formal) power series defined by

$$G(z) = Z\{g_k\} = \sum_{k=0}^\infty g_k z^{-k} \ ,$$

where z is the complex variable. Again the most important property for our purpose is that it changes a difference equation to an algebraic equation via

$$Z\{g_{k+1}\} = z\{Z\{g_k\} - g_0\} \ , \tag{5.8}$$

etc. It is important to observe that (5.7 and 8) are *completely analogous*. Hence, it is sufficient to consider the continuous-time setting. For convenience, we will also assume that the initial state is 0. Hence, taking the Laplace transform of each term in the state-space description

$$\dot{x} = Ax + Bu$$

$$v = Cx$$

where A, B, C are of course constant $n \times n$, $n \times p$ and $q \times n$ matrices, we have

$$sX = AX + BU$$
$$V = CX \tag{5.9}$$

which yields the input-output relationship

$$V = H(s)U \ , \tag{5.10}$$

where $H(s)$, called the *transfer function* of the linear system, is defined by

$$H(s) = C(sI - A)^{-1}B \ .$$

Here, it is clear that, at least for large values of s, $sI - A$ is invertible, and its inverse is an analytic function of s and hence can be continued analytically to the entire complex s-plane with the exception of at most n poles which are introduced by the zeros of the nth degree polynomial $\det(sI - A)$. In fact, if we use the notation

$$(sI - A)^*$$

to denote the $n \times n$ matrix whose (i, j)th entry is $(-1)^{i+j} \det \hat{A}_{ij}(s)$, where $\hat{A}_{ij}(s)$ is the $(n-1) \times (n-1)$ sub-matrix of $sI - A$ obtained by deleting the jth row and ith column, we have

$$H(s) = \frac{C(sI - A)^*B}{\det(sI - A)} \ . \tag{5.11}$$

Here, the numerator is a $q \times p$ matrix, each of whose entries is a polynomial in s of degree at most $n-1$, and the denominator is a (scalar-valued) nth degree polynomial with leading coefficient 1. It is possible that a zero of the denominator cancels with a common zero of the numerator.

5.4 Pole-Zero Cancellation of Transfer Functions

An important problem in linear system theory is to obtain a state-space description of the linear system from its transfer function $H(s)$, so that the state vector has the lowest dimension. This is called the problem of *minimal realization* (Sect. 10.5). To achieve a minimal realization it is important to reduce the denominator in (5.11) to its lowest degree. This reduction is called *pole-zero cancellation*.

Definition 5.1   The transfer function $H(s)$ is said to have no pole-zero cancellation if none of the zeros of the denominator $\det(sI - A)$ in (5.11) disappears by all possible cancellations with the numerator, although there might be some reduction in the orders of these zeros.

It is quite possible to have a pole-zero cancellation as can be seen in the following example. Consider

$$A = \begin{bmatrix} -2 & 1 \\ 3 & 0 \end{bmatrix}, \quad B = \begin{bmatrix} 1 \\ -1 \end{bmatrix}, \quad C = \begin{bmatrix} 0 & -1 \end{bmatrix} . \tag{5.12}$$

Then the transfer function of the state-space description defined by these matrices is

$$H(s) = \frac{\begin{bmatrix} 0 & -1 \end{bmatrix} \begin{bmatrix} s & 1 \\ 3 & s+2 \end{bmatrix} \begin{bmatrix} 1 \\ -1 \end{bmatrix}}{\det \begin{bmatrix} s+2 & -1 \\ -3 & s \end{bmatrix}} = \frac{(s-1)}{(s+3)(s-1)} .$$

Hence, the zero $s = 1$ in the denominator [i.e. the possible "pole" of $H(s)$] cancels with the numerator. This pole-zero cancellation makes $H(s)$ analytic on the right-half complex s-plane as well as on the imaginary axis, which is usually used as a test for stability (Chap. 6). It will be seen in Chap. 6, however, that this system is not state-stable although it is input-output stable. Hence, an important information on instability, namely that $s = 1$ being an eigenvalue of A, is lost. This does not occur for completely controllable and observable linear systems.

Theorem 5.2 *The transfer function $H(s)$ of the state-space description*

$$\dot{x} = Ax + Bu$$

$$v = Cx$$

of a time-invariant linear system which is both completely controllable and observable has no pole-zero cancellation in the expression (5.11).

The proof of this theorem depends on some properties of minimum polynomials, for which we refer the reader to a book on linear algebra; see, for example, Nering (1963). Recall that the minimum polynomial $q_m(s)$ of the $n \times n$ system matrix A is the lowest degree polynomial with leading coefficient 1, such that $q_m(A) = 0$. Hence, $m \le n$ and, in fact, if $d(s)$ is the greatest common divisor, again with leading coefficient 1, of *all* the entries of $(sI - A)^*$, then

$$q_m(s) = \frac{\det(sI - A)}{d(s)} . \tag{5.13}$$

Let us define a matrix $F(s)$ by $(sI - A)^* = d(s)F(s)$. Then we have $d(s)(sI - A)F(s) = (sI - A)(sI - A)^* = \det(sI - A)I$, so that

$$q_m(s)I = (sI - A)F(s) , \tag{5.14}$$

and taking the determinant of both sides yields

$$q_m^n(s) = \det(sI - A) \det F(s) .$$

This shows the important property that the zeros of the characteristic polynomial $\det(sI - A)$ are also the zeros of the minimum polynomial $q_m(s)$. On the other hand, we have

$$H(s) = \frac{C(sI - A)^* B}{\det(sI - A)} = \frac{d(s)CF(s)B}{d(s)q_m(s)} = \frac{CF(s)B}{q_m(s)}$$

by using (5.11, 13), and the definition of $F(s)$. Hence, to show that there is no pole-zero cancellation, it is sufficient to show that if $q_m(s^*) = 0$ then $CF(s^*)B$ is not the zero $q \times p$ matrix.

To prove this assertion, we need more information on $F(s)$. Write

$$q_m(s) = s^m - a_1 s^{m-1} - \ldots - a_m .$$

It can be shown (Exercise 5.12) that

$$q_m(s) - q_m(t) = (s - t) \sum_{k=0}^{m-1} (t^k - a_1 t^{k-1} - \ldots - a_k)s^{m-k-1} . \tag{5.15}$$

Hence, replacing s and t by the matrices sI and A, respectively, and noting that $q_m(A) = 0$, we have

$$q_m(s)I = q_m(sI) = (sI - A) \sum_{k=0}^{m-1} (A^k - a_1 A^{k-1} - \ldots - a_k I)s^{m-k-1} .$$

This together with (5.14) gives

$$F(s) = \sum_{k=0}^{m-1} (A^k - a_1 A^{k-1} - \ldots - a_k I)s^{m-k-1} . \tag{5.16}$$

As a consequence of (5.16), we observe that $F(s)$ commutes with any power of A, i.e.

$$F(s)A^k = A^k F(s), \quad k = 1, 2, \ldots . \tag{5.17}$$

Assume, on the contrary, that both $q_m(s^*) = 0$ and $CF(s^*)B = 0$. Then by (5.14), we have

$$(s^*I - A)F(s^*) = q_m(s^*)I = 0$$

so that $AF(s^*) = s^* F(s^*)$, and $A^2 F(s^*) = s^* AF(s^*) = s^{*2} F(s^*)$, etc. Hence, from (5.17) we have

$$A^k F(s^*) = F(s^*)A^k = s^{*k} F(s^*), \quad k = 1, 2, \ldots . \tag{5.18}$$

One consequence is that

$$CA^k F(s^*)B = s^{*k}[CF(s^*)B] = 0, \quad k = 0, 1, \ldots ,$$

or $N_{CA}(F(s^*)B) = 0$, where N_{CA} is the observability matrix. Since the linear system is observable, the column rank of N_{CA} is full, and this implies that $F(s^*)B = 0$. We can now apply (5.17) to obtain

$$F(s^*)A^k B = A^k F(s^*)B = 0, \quad k = 0, 1, \ldots ,$$

or $F(\dot{s}^*)M_{AB} = 0$, where M_{AB} is the controllability matrix. Since the linear system is completely controllable, the rank of M_{AB} is full, so that $F(s^*) = 0$. If $s^* = 0$, then (5.16) gives

$$A^{m-1} - a_1 A^{m-2} - \ldots - a_{m-1}I = 0$$

which contradicts that the minimum polynomial $q_m(s)$ is of degree m, and if $s^* \neq 0$, then again by (5.16),

$$p(A) = 0 \quad \text{where}$$

$$p(s) = \sum_{k=0}^{m-1} (s^k - a_1 s^{k-1} - \ldots - a_k)s^{*m-k-1}$$

is a polynomial of degree $m-1$, and we also arrive at the same contradiction. This completes the proof of the theorem.

Exercises

5.1 Give some examples to convince yourself of the statement made in Remark 5.1. Then prove that this statement holds in general.

5.2 If a state-space description of a continuous-time linear system with zero transfer matrix is completely controllable, show that the same description with a nonzero transfer matrix $D(t)$ is also completely controllable. Repeat the same problem for observability.

5.3 Show that an additional free vector $f(t)$ in (5.1) does not change the controllability and observability of the linear system.
(*Hint*: Return to the definitions).

5.4 Formulate and justify Remarks 1, 2, and 3 for discrete-time linear systems.

5.5 Complete the proof of Lemma 5.1 by showing that V_4 is an invariant subspace of $\mathbb{R}^n$ under A^T.

5.6 In the example described by (5.3), show that no unitary transformation W can make $\mathscr{S}_1$ controllable without changing the desired decomposed form.

5.7 Verify that the subsystem $\mathscr{S}_2$ in the example described by (5.3) is completely controllable and observable; that the combined subsystem of $\mathscr{S}_1$ and $\mathscr{S}_2$ is completely controllable; that the combined subsystem $\mathscr{S}_2$ and $\mathscr{S}_4$ is observable; that the subsystem $\mathscr{S}_4$ is observable; but that the subsystem $\mathscr{S}_1$ is *not* controllable. Also, verify that if the transformation G^{-1} is used, where G is given by (5.5), then the transformed subsystem $\mathscr{S}_1$ is now completely controllable while $\mathscr{S}_2$, $\mathscr{S}_3$, and $\mathscr{S}_4$ remain unchanged.

5.8 Prove that the dimensions $n_1, \ldots, n_4$ of the subsystems in the decomposed system (Theorem 5.1) are invariant under nonsingular transformations.

5.9 Verify that the combined subsystem $\mathscr{S}_1$ and $\mathscr{S}_2$ in Theorem 5.1 is completely controllable, and the combined subsystem $\mathscr{S}_2$ and $\mathscr{S}_4$ is observable. Complete the proof of Theorem 5.1 by verifying that the appropriate controllability and observability matrices are of full rank.

5.10 Verify the z-transform property (5.8) and generalize to $Z\{g_{k+j}\}$.

5.11 Verify that there is a pole-zero cancellation in the example (5.12), and determine the ranks of the controllability and observability matrices.

5.12 Derive the formula given by (5.15).

5.13 (a) If

$$A = \begin{bmatrix} 1 & 0 \\ 0 & 1 \end{bmatrix} \quad \text{and} \quad B = \begin{bmatrix} 1 \\ 1 \end{bmatrix},$$

verify that the system is not controllable while the two subsystems $\mathscr{S}_1$ and $\mathscr{S}_2$ are completely controllable.

(b) If

$$A = \begin{bmatrix} 1 & 1 \\ 0 & 1 \end{bmatrix} \quad \text{and} \quad B = \begin{bmatrix} 0 \\ 1 \end{bmatrix},$$

verify that the system and its subsystem $\mathscr{S}_2$ are both completely controllable while $\mathscr{S}_1$ is not.

6. Stability

The origin of the notion of stability dates back to the 1893 paper of A. M. Lyapunov, entitled "Problème général de la stabilité du mouvement". In this chapter we only discuss the stability of linear systems. As usual, we begin with the continuous-time setting.

6.1 Free Systems and Equilibrium Points

A system with zero input is called a *free system*. Hence, a free linear system can be described by

$$\dot{x} = A(t)x , \tag{6.1}$$

where the entries of the $n \times n$ system matrix $A(t)$ will be assumed, as usual, to be continuous functions on an interval J that extends to $+\infty$. A position x_e in $\mathbb{R}^n$ is called an *equilibrium point* (or *state*) of the system described by (6.1) if the initial-value problem

$$\dot{x} = A(t)x, \quad t \geq t_0$$
$$x(t_0) = x_e$$

has the unique solution $x(t) = x_e$ for all $t \geq t_0$. This, of course, means that with x_e as the initial state there is absolutely no movement at all. For instance, any position $[a \ 0]^T$, where a is arbitrarily chosen, is an equilibrium point of the free system

$$\dot{x} = \begin{bmatrix} 0 & 1 \\ 0 & 0 \end{bmatrix} x .$$

More generally, if $\Phi(t, t_0)$ denotes the transition matrix of (6.1), then x_e is an equilibrium point if and only if

$$[I - \Phi(t, t_0)]x_e = 0$$

for all $t \geq t_0$. Hence, if the matrix $I - \Phi(t, t_0)$ is nonsingular for some $t > t_0$, then the only equilibrium point is the origin.

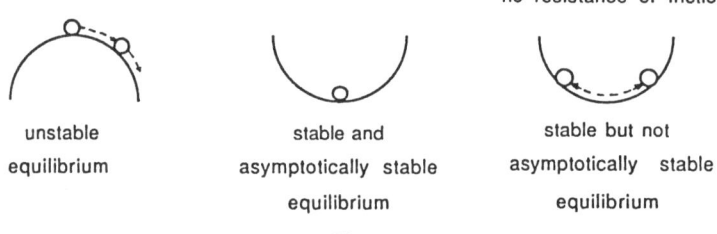

no resistance or friction

unstable   stable and   stable but not
equilibrium   asymptotically stable   asymptotically stable
                   equilibrium   equilibrium

Fig. 6.1

It is interesting to study how the state vector behaves if the initial state is near but not at an equilibrium point. A ball sitting still on top of a hill will roll away when it is disturbed, but if it is slightly perturbed while sitting on the bottom of a valley, it will eventually move back to the original equilibrium position. However, if there is no resistance, the perturbed ball on the bottom of a frictionless valley just oscillates back and forth, but never stays at the bottom. These phenomena illustrate the notion of unstable equilibrium, asymptotically stable equilibrium, and stable equilibrium in the sense of Lyapunov, respectively (Fig. 6.1).

6.2 State-Stability of Continuous-Time Linear Systems

In this section, we introduce three related but different types of state-stability.

Definition 6.1 A free linear system described by (6.1) is said to be *stable (in the sense of Lyapunov)* about an equilibrium point x_e (or equivalently, x_e is a stable equilibrium point of the system) if for any $\varepsilon > 0$, there exists a $\delta > 0$, such that $|x(t) = x_e| < \varepsilon$ for all sufficiently large t whenever $|x(t_0) = x_e|_2 < \delta$ (cf. Exercise 2.6 for definition of the "length" $| \ |_2$ and Remark 6.3 below).

Another terminology for stablity in the sense of Lyapunov is *state-stability*, since it describes the stability of the state vector.

Definition 6.2 A free linear system is said to be *unstable* about an equilibrium point x_e (or x_e is an unstable equilibrium point of the system) if it is not stable about x_e; that is, there exists an $\varepsilon_0 > 0$ such that for every $\delta > 0$, some initial state $x(t_0)$ and a sequence $t_k \to +\infty$ can be chosen to satisfy $|x(t_0) - x_e|_2 < \delta$ and $|x(t_k) - x_e|_2 \geq \varepsilon_0$ for all k.

Definition 6.3 A free linear system is said to be *asymptotically stable* about an equilibrium point x_e (or x_e is an asymptotically stable equilibrium point of the system) if there exists a $\delta > 0$ such that $|x(t) - x_e|_2 \to 0$ as $t \to +\infty$ whenever $|x(t_0) - x_e|_2 < \delta$.

This stability is also called *asymptotic state-stability*.

Clearly, an asymptotically stable equilibrium point is also a stable equilibrium point in the sense of Lyapunov, but the converse is false as illustrated in the frictionless valley example. More precisely, the free linear system

$$\dot{x} = \begin{bmatrix} 0 & 1 \\ -1 & 0 \end{bmatrix} x$$

has $x_e = 0$ as an equilibrium point, and if $x(t_0) = [\delta_1 \ \delta_2]^T$ where $\delta_1^2 + \delta_2^2 > 0$, then it can be seen that

$$x(t) = [\delta_1 \cos(t-t_0) + \delta_2 \sin(t-t_0) \quad -\delta_1 \sin(t-t_0) + \delta_2 \cos(t-t_0)]^T$$

for all $t \geq t_0$ so that $|x(t) - x_e|_2 = |x(t)|_2 = |x(t_0)|_2$ for all t (and we could have chosen δ to be the given ε), but that $x(t)$ clearly does not converge to 0.

Remark 6.1   Using the translation $y = x - x_e$ we may (and will) assume that the equilibrium point is 0. The system description is unchanged under this translation since

$$\dot{y} = \frac{d}{dt}(x - \Phi(t, t_0)x_e)$$

$$= \dot{x} - \frac{d}{dt}\Phi(t, t_0)x_e$$

$$= A(t)x - A(t)\Phi(t, t_0)x_e$$

$$= A(t)x - A(t)x_e$$

$$= A(t)(x - x_e) = A(t)y$$

which is the same equation (6.1) that x satisfies.

Remark 6.2   The restriction of $|x(t_0)|_2 < \delta$ in the definition of asymptotic stability can be omitted for free linear systems, since $\delta x(t) = \Phi(t, t_0)[\delta x(t_0)]$ and $|x(t)|_2 \to 0$ if and only if $\delta|x(t_0)|_2 \to 0$ as $t \to +\infty$.

Remark 6.3   If $x = [x_1 \ldots x_n]^T$, then

$$|x|_2 = \sqrt{x_1^2 + \ldots + x_n^2}$$

is the actual length of x in $\mathbb{R}^n$. This generalizes to $|F|_2$ of a matrix $F = [f_{ij}]$ by defining

$$|F|_2 = \left(\sum_{i,j} f_{ij}^2 \right)^{1/2} .$$

For convenience, we will sometimes drop the subscript 2, so that $|x| = |x|_2$ and $|F| = |F|_2$.

Theorem 6.1 *Let $\Phi(t, t_0)$ be the transition matrix of the free linear system described by (6.1). This system is stable about 0 if and only if there exists some positive constant C, depending only on t_0, such that*

$$|\Phi(t, t_0)| \le C \tag{6.2}$$

for all $t \ge t_0$. It is asymptotically stable about 0 if and only if

$$|\Phi(t, t_0)| \to 0 \tag{6.3}$$

as $t \to +\infty$.

Recall that $x(t) = \Phi(t, t_0)x(t_0)$ since we have zero control function $\boldsymbol{u}$. By Schwarz's inequality, we obtain

$$|x(t)| \le |\Phi(t, t_0)| |x(t_0)| \tag{6.4}$$

(Exercise 6.4). Hence, if (6.2) is satisfied, then for a given $\varepsilon > 0$, we can choose $\delta = \varepsilon/C$, so that the system is stable about 0. Furthermore, if (6.3) is satisfied, then the above inequality gives $|x(t)| \to 0$, so that the system is asymptotically stable about 0.

To see the converse of the first statement, we assume that the system is stable about 0 but, on the contrary, (6.2) is not satisfied for any C. That is, there is some entry $\phi_{i_0, j_0}(t, t_0)$ in $\Phi(t, t_0)$, $1 \le i_0, j_0 \le n$, that is unbounded, as $t \to +\infty$. Let $x(t_0) = [0 \ldots 0 \ 1 \ 0 \ldots 0]^T$, 1 being placed in the j_0th entry. Then $|x(t)| = |\Phi(t, t_0)x(t_0)| \ge |\phi_{i_0 j_0}(t, t_0)|$ which is unbounded (cf. Remark 6.2 for dropping the requirement $|x(t_0)| < \delta$), contradicting the stability assumption. The proof of the converse of the second statement is similar (Exercise 6.7). This completes the proof of the theorem.

Let us consider time-invariant systems for the time being and denote by $\lambda_j = r_j + is_j, (r_j, s_j$ real$) j = 1, \ldots, k$, the eigenvalues of the $n \times n$ constant matrix A with multiplicities $m_1, \ldots, m_k$, respectively $(m_1 + \ldots + m_k = n)$, so arranged that $r_1 \ge r_2 \ge \ldots \ge r_k$. Now if $\Phi(t, 0)$ is the transition matrix with initial time $t_0 = 0$, its Laplace transform is

$$(\mathscr{L}\Phi)(s) = \left(\sum_{j=0}^{\infty} \frac{t^j}{j!} A^j \right) = \sum_{j=0}^{\infty} \frac{1}{s^{j+1}} A^j ,$$

so that

$$(sI - A)(\mathscr{L}\Phi)(s) = \sum_{j=0}^{\infty} \frac{1}{s^j} A^j - \sum_{j=0}^{\infty} \frac{1}{s^{j+1}} A^{j+1} = I , \qquad \text{or}$$

$$\begin{aligned} (\mathscr{L}\Phi)(s) &= (sI - A)^{-1} \\ &= \frac{(sI - A)^*}{\det(sI - A)} \\ &= \frac{(sI - A)^*}{\prod\limits_{j=1}^{k} (s - \lambda_j)^{m_j}} . \end{aligned}$$

Since each entry in $(sI - A)^*$ is a polynomial of degree $< n$ and the denominator is of degree $= n$, we can use partial fractions and obtain

$$(\mathscr{L}\,\Phi)(s) = \sum_{j=1}^{k} \sum_{l=0}^{m_j-1} \frac{P_{lj}}{(s-\lambda_j)^{l+1}} \;,$$

where P_{lj} are $n \times n$ constant matrices (with complex entries). Taking the inverse Laplace transformation, we have

$$\Phi(t, 0) = e^{tA} = \sum_{j=1}^{k} \sum_{l=0}^{m_j-1} \frac{t^l}{l!}\, e^{\lambda_j t}\, P_{lj}\;.$$

Hence, the transition matrix corresponding to a given constant matrix A and with initial time t_0 has the following expression:

$$\Phi(t, t_0) = e^{(t-t_0)A} = \sum_{j=1}^{k} \sum_{l=0}^{m_j-1} \frac{(t-t_0)^l}{l!}\, e^{\lambda_j(t-t_0)}\, P_{lj}\;. \tag{6.5}$$

This formulation of $\Phi(t, t_0)$ is very useful in the study of stability. For instance, if we write $r_1 = \ldots = r_p > r_{p+1} \geq \ldots \geq r_k$ $(p \geq 1)$ and set $r = r_1$, then (6.5) yields

$$|\Phi(t, t_0)| = e^{r(t-t_0)} \left| \sum_{j=1}^{p} e^{is_j(t-t_0)} \sum_{l=0}^{m_j-1} \frac{(t-t_0)^l}{l!}\, P_{lj} + o(1) \right| \tag{6.6}$$

where o(1) (which reads "small 'oh' one") is a so-called *Landau notation* that denotes the error term that tends to 0 as $t \to +\infty$. The following result is a simple consequence of this estimate and Theorem 6.1 (Exercises 6.8 and 10).

Theorem 6.2 *Let the time-invariant system matrix A in (6.1) be an $n \times n$ matrix with eigenvalues λ_j. Then the corresponding continuous-time free linear system is asymptotically stable about 0 if and only if $\mathrm{Re}\{\lambda_j\} < 0$ for all j. It is stable about 0 in the sense of Lyapunov if and only if $\mathrm{Re}\{\lambda_j\} \leq 0$ for all j, and for each j with $\mathrm{Re}\{\lambda_j\} = 0$, λ_j is a simple eigenvalue of A.*

Remark 6.4 The result in the above theorem does not apply to time-varying systems. For example, if

$$A(t) = \begin{bmatrix} -4 & 3e^{-8t} \\ -e^{8t} & 0 \end{bmatrix},$$

the eigenvalues of $A(t)$ are $\lambda = -1$ and -3 (independent of t) which of course have negative real parts. However, with the initial state $x(0) = [\delta \quad \delta]^T$, $\delta > 0$, the state vector becomes

$$x(t) = \begin{bmatrix} (3e^{-5t} - 2e^{-7t})\delta \\ (2e^t - e^{3t})\delta \end{bmatrix}$$

so that $|x(t)| \to \infty$ as $t \to +\infty$ for any $\delta > 0$, no matter how small. That is, this system is even unstable about 0.

Remark 6.5 Let the time-invariant system described in Theorem 6.2 be asymptotically stable. Then all the eigenvalues λ_j of the system matrix A have negative real parts. Choose any ρ that satisfies

$$0 < \rho < \min(-\text{Re}\{\lambda_j\}) \ .$$

Then the estimate (6.6) gives

$$|\Phi(t, t_0)| \le e^{-\rho(t-t_0)}$$

for all large values of t (Exercise 6.9). In particular, if x is the state vector with initial state $x(t_0)$, then x satisfies

$$|x(t)| \le |x(t_0)|e^{-\rho(t-t_0)} \ . \tag{6.7}$$

This shows that not only does $|x(t)|$ tend to 0, it tends to 0 exponentially fast.

Time-varying systems, however, do not necessarily have this property as can be seen from the example $\dot{x}(t) = -t^{-1}x(t)$ where $t \ge t_0 > 0$, since the solution of this initial-value problem is

$$x(t) = x(t_0)(t_0 t^{-1})$$

which tends to zero as $t \to +\infty$, but certainly does not tend to zero exponentially fast as (6.7). So for time-varying systems, we need the following finer stability classification.

Definition 6.4 A free linear system described by (6.1) is said to be *exponentially stable* about the equilibrium point 0, if there exists a positive constant ρ such that the state vector $x(t)$ satisfies the inequality (6.7) for all sufficiently large values of t and any initial state $x(t_0)$. (Note that in view of Remark 6.2, we have no longer required $|x(t_0)| < \delta$.)

The following result characterizes all such free linear systems.

Theorem 6.3 *Let $|A(t)| \le M_0 < \infty$ for all $t \ge t_0$. Then the corresponding free linear system is exponentially stable about the equilibrium point 0 if and only if there exists a positive constant M_1 such that the transition matrix $\Phi(t, s)$ of A with initial time $s \ge t_0$ satisfies*

$$\int_s^t |\Phi(\tau, s)| d\tau \le M_1 < \infty \tag{6.8}$$

for all $t \ge s \ge t_0$.

One direction of this theorem is intuitively clear since state vectors and the transition matrix are intimately related. In fact, the jth column $\phi_j = \phi_j(t, s)$ of

$\Phi(t, s)$ is the state vector $x(t)$ with initial state $x(s) = [0 \ldots 0 \ 1 \ 0 \ldots 0]^T$, 1 being placed at the jth component. Hence, if the system is exponentially stable about 0, then $\rho > 0$ exists such that

$$|\phi_j(t, s)| \le e^{-\rho(t-s)}$$

for all sufficiently large values of t, and $j = 1, \ldots, n$. This gives

$$\int_s^t |\Phi(\tau, s)| d\tau = \int_s^t \left\{ \sum_{j=1}^n |\phi_j(\tau, s)|^2 \right\}^{1/2} d\tau \le n^{1/2} \int_s^t e^{-\rho(\tau-s)} d\tau < \frac{n^{1/2}}{\rho}$$

for all $t \ge s \ge t_0$. To prove the converse, assume that $\Phi(t, s)$ satisfies (6.8). Our first observation is that $|\Phi(t, s)|$ is uniformly bounded for all t and s with $t \ge s$. Indeed, if $t \ge s$, we have

$$|\Phi(t, s) - I| = \left| \int_s^t \frac{\partial}{\partial w} \Phi(w, s) \, dw \right|$$

$$= \left| \int_s^t A(w) \Phi(w, s) \, dw \right|$$

$$\le \int_s^t |A(w) \Phi(w, s)| \, dw$$

$$\le \int_s^t |A(w)| \, |\Phi(w, s)| \, dw \le M_0 M_1 \ ,$$

where Schwarz's inequality and the inequality in Exercise 6.6 have been used, and hence an application of the triangle inequality (Exercise 6.5) gives

$$|\Phi(t, s)| \le |I| + M_0 M_1 = n^{1/2} + M_0 M_1 := M_2 \ ,$$

say. Next, by using a property of the transition matrix, we have, for $t \ge s \ge t_0$,

$$(t - s)|\Phi(t, s)| = \int_s^t |\Phi(t, s)| \, dw$$

$$= \int_s^t |\Phi(t, w) \Phi(w, s)| \, dw$$

$$\le \int_s^t |\Phi(t, w)| \, |\Phi(w, s)| \, dw$$

$$\le M_2 \int_s^t |\Phi(w, s)| \, dw \le M_1 M_2 \ .$$

Hence, whenever $(t-s) \geq 2M_1 M_2$, we have

$$|\Phi(t,s)| \leq \tfrac{1}{2} .\tag{6.9}$$

Now, starting at t_0, if $t > t_0$ we can choose the largest nonnegative integer k satisfying $t_0 + k(2M_1 M_2) \leq t$ so that $k > [(t-t_0)/2M_1 M_2] - 1$, and using the notation

$$t_k = t_0 + 2kM_1 M_2 ,$$

we obtain

$$|\Phi(t,t_0)| = |\Phi(t,t_{k-1})\Phi(t_{k-1},t_{k-2}) \ldots \Phi(t_1,t_0)|$$

$$\leq |\Phi(t,t_{k-1})||\Phi(t_{k-1},t_{k-2})| \ldots |\Phi(t_1,t_0)|$$

$$\leq 2^{-k} < 2e^{-r(t-t_0)}$$

by defining $r = (\ln 2)/(2M_1 M_2)$. Here, we have used Schwarz's inequality and inequality (6.9) $(k-1)$ and k times, respectively, and of course, the last inequality follows from the definition of k. Hence, again by using Schwarz's inequaltiy, we obtain

$$|x(t)| = |\Phi(t,t_0)x(t_0)| \leq |\Phi(t,t_0)| \, |x(t_0)| \leq 2e^{-r(t-t_0)}|x(t_0)|$$

which gives (6.7) for all large values of t by choosing any ρ with $0 < \rho < r$. This completes the proof of the theorem.

6.3 State-Stability of Discrete-Time Linear Systems

We now turn to the study of the discrete-time setting. To do so, we need a result from linear algebra. Recall that any $n \times n$ constant matrix A is similar to a Jordan canonical form J; that is $A = PJP^{-1}$ for some nonsingular matrix P. The reader probably remembers that J has at most two nonzero diagonals; namely, the main diagonal that consists of all eigenvalues of A listed according to their multiplicities, and the one above the main diagonal that consists of only 0 or 1. For our purpose in studying the stability of discrete-time systems, we have to be more precise. Let $\lambda_1, \ldots, \lambda_l$ be the distinct eigenvalues of A, and let the characteristic and minimum polynomials of A be

$$\det(sI - A) = (s - \lambda_1)^{n_1} \ldots (s - \lambda_l)^{n_l} \quad \text{and}$$

$$q(s) = (s - \lambda_1)^{m_1} \ldots (s - \lambda_l)^{m_l} ,$$

respectively, where $n_1 + \ldots + n_l = n$ and $m_i \leq n_i, i = 1, \ldots, l$. For each i, let s_i be the dimension of the vector space spanned by all eigenvectors corresponding to the eigenvalue λ_i. s_i is called the *geometric multiplicity* of λ_i and n_i is called the *algebraic multiplicity* of λ_i. It is known that $s_i \leq n_i$ also. To understand the Jordan canonical form J, it is best to imagine J as a block diagonal matrix. It turns out that the number of diagonal blocks that contain λ_i on their main diagonals is equal to s_i. In fact, we can write

$$
J = \begin{bmatrix} B_1(\lambda_1) & & & & \\ & \ddots & & & \\ & & B_{s_1}(\lambda_1) & & \\ & & & \ddots & \\ & & & & B_1(\lambda_l) \\ & & & & & \ddots \\ & & & & & & B_{s_l}(\lambda_l) \end{bmatrix}
$$

where the blocks that are not listed are zero blocks and

$$
B_i(\lambda_j) = \begin{bmatrix} \lambda_j & 1 & & \\ & \ddots & \ddots & \\ & & \ddots & 1 \\ & & & \lambda_j \end{bmatrix},
$$

$i = 1, \ldots, s_j$ and $j = 1, \ldots, l$, and again the entries that are not listed are zeros. In addition, for each $j = 1, \ldots, l$, the "leading block" $B_1(\lambda_j)$ is an $m_j \times m_j$ submatrix while the orders of the other blocks $B_i(\lambda_j)$, $i > 1$, are less than or equal to m_j, such that the sum of the orders of all $B_1(\lambda_j), \ldots, B_{s_j}(\lambda_j)$ is exactly n_j. It is also known that with the exception of a permutation of the diagonal blocks, the Jordan canonical form J of A is unique.

One important consequence is that if $m_i = 1$, then the $n_i \times n_i$ submatrix A_i of J, consisting of the totality of all blocks that contain the eigenvalue λ_i, is a diagonal submatrix; that is,

$$
A_i = \begin{bmatrix} B_1(\lambda_1) & & \\ & \ddots & \\ & & B_{s_i}(\lambda_i) \end{bmatrix} = \begin{bmatrix} \lambda_i & 0 & \cdots & \cdots & 0 \\ 0 & \ddots & & & \vdots \\ \vdots & & \ddots & & \vdots \\ \vdots & & & \ddots & 0 \\ 0 & \cdots & \cdots & 0 & \lambda_i \end{bmatrix} = \lambda_i I . \tag{6.10}
$$

Another important consequence is that if $m_j \geq 2$, then there is at least a 1 on the $(i, i+1)$ diagonal of the corresponding $n_j \times n_j$ submatrix A. More precisely,

$$
A_j = \begin{bmatrix} B_1(\lambda_j) & & \\ & \ddots & \\ & & B_{s_j}(\lambda_j) \end{bmatrix} = \begin{bmatrix} \lambda_j & 1 & 0 & \cdots & \cdots & 0 \\ 0 & \lambda_j & b_1 & & & \vdots \\ \vdots & & \ddots & \ddots & & 0 \\ \vdots & & & \ddots & \ddots & b_{n_j-2} \\ 0 & \cdots & \cdots & \cdots & 0 & \lambda_j \end{bmatrix} \tag{6.11}
$$

where $b_1, \ldots, b_{n_j-2}$ are 0 or 1.

We can now discuss the problem of stability for discrete-time linear systems. Let A be an $n \times n$ constant matrix and consider the time-invariant free linear system

$$x_{k+1} = Ax_k \ . \tag{6.12}$$

Without loss of generality, we assume in the following discussion that the initial time is $k=0$ so that the initial state is x_0.

Definition 6.5   A discrete-time free linear system described by (6.12) is said to be *stable (in the sense of Lyapunov)* about 0 if for any $\varepsilon > 0$, there exists a $\delta > 0$ such that $|x_k| < \varepsilon$ for all sufficiently large values of k whenever $|x_0| < \delta$. It is said to be *asymptotically stable* about 0, if $|x_k| \to 0$ as $k \to \infty$, or equivalently,

$$\lim_{k \to \infty} |A^k x_0| = 0 \tag{6.13}$$

for all x_0 in $\mathbb{R}^n$. (Note that in view of Remark 6.2, we have dropped the requirement $|x_0| < \delta$ in the definition of asymptotic stability about 0.)

Again asymptotic stability is a stronger notion than stability in the sense of Lyapunov. In fact we have the following characterization.

Theorem 6.4   *Let $\lambda_j, j = 1, \ldots, l$, be the distinct eigenvalues of the $n \times n$ matrix A. Then the corresponding discrete-time free linear system (6.12) is asymptotically stable about 0 if and only if $|\lambda_j| < 1, j = 1, \ldots, l$. It is stable about 0 in the sense of Lyapunov if and only if $|\lambda_j| \leq 1$ for all j, and for each j with $|\lambda_j| = 1$, λ_j is a simple root of the minimum polynomial $q(s)$ of A.*

Our proof of this theorem relies on the Jordan canonical form J of A as discussed early. We do not, however, require the fine structure of the diagonal blocks $B_j(\lambda_i)$ but only the weaker diagonal blocks A_i as given in (6.10, 11). Let us arrange the eigenvalues $\lambda_1, \ldots, \lambda_l$ in such a way that $m_1 = \ldots = m_p = 1$ and $m_{p+1}, \ldots, m_l > 1$. Then we have from (6.10, 11).

$$P^{-1}AP = J = \begin{bmatrix} \lambda_1 I & & & & & \\ & \ddots & & & & \\ & & \lambda_p I & & & \\ & & & A_{p+1} & & \\ & & & & \ddots & \\ & & & & & A_l \end{bmatrix}$$

where, for $j = p+1, \ldots, l$, A_j is an $n_j \times n_j$ submatrix $(n_j \geq m_j \geq 2)$ given by (6.11).

Hence, taking the kth power, we have

$$P^{-1}A^kP=J^k=\begin{bmatrix} \lambda_1^k I & & & & & \\ & \ddots & & & & \\ & & \lambda_p^k I & & & \\ & & & A_{p+1}^k & & \\ & & & & \ddots & \\ & & & & & A_l^k \end{bmatrix}\qquad(6.14)$$

with

$$A_j^k=\begin{bmatrix} \lambda_j^k & k\lambda_j^{k-1} & * & \cdots & * \\ 0 & \lambda_j^k & * & \cdots & * \\ \vdots & & \ddots & \ddots & \vdots \\ \vdots & & & \ddots & * \\ 0 & \cdots\cdots\cdots & 0 & & \lambda_j^k \end{bmatrix}\qquad(6.15)$$

where each $*$ denotes a term whose magnitude is bounded by

$$k\ldots(k-i+1)\lambda_j^{k-i},\quad 1\le i\le n_j-1,\quad j=p+1,\ldots,l\ .$$

First we note that if all $|\lambda_j|<1$, then $|x_k|=|A^k x_0|=|PJ^kP^{-1}x_0|$ $\le|P|\,|P^{-1}x_0||J^k|$ which tends to 0, since each entry of J^k tends to 0 as $k\to\infty$ (Exercise 6.8). Conversely, if (6.13) holds, then we must also have $|\lambda_j|<1$ for all j, since by choosing

$$x_0=P[\underbrace{1\ 0\ldots0}_{n_1}\ \underbrace{1\ 0\ldots0}_{n_2}\ldots\underbrace{1\ 0\ldots0}_{n_l}]^T\ ,$$

it follows from (6.14) and (6.15) that

$$(|\lambda_1|^{2k}+\ldots+|\lambda_l|^{2k})^{1/2}=|J^kP^{-1}x_0|=|P^{-1}A^kx_0|\le|P^{-1}|\,|A^kx_0|\ .$$

To establish the second statement in Theorem 6.4, we first assume that if $|\lambda_j|$ $=1$ then m_j is 1, so that $1\le j\le p$, and consequently $|\lambda_i|<1$ for $i=p+1,\ldots,l$. Hence, for each x_0, writing

$$x_0=P[y_1\cdots y_{n_1+\ldots+n_p}\ \ y_{n_1+\ldots+n_p+1}\cdots y_n]^T\ ,$$

we have

$$|x_k|=|A^kx_0|=|PJ^kP^{-1}x_0|\le|P|\,|J^kP^{-1}x_0|$$
$$=|P|\{(y_1^2+\ldots+y_{n_1+\ldots+n_p}^2)^{1/2}+o(1)\}\ ,$$

where the o(1) term is a contribution from the eigenvalues λ_i, $i \geq p+1$, and this term tends to zero since $|\lambda_i| < 1$ (Exercise 6.8). Hence, for every given $\varepsilon > 0$, we can find a $\delta > 0$ to control the term $y_1^2 + \ldots + y_{n_1}^2 + \ldots + n_p$, so that $|x_0| < \delta$ implies $|x_k| < \varepsilon$ for all large values of k. Conversely, if $|\lambda_j| = 1$ but λ_j is not a simple root of the minimum polynomial of A, i.e. $m_j \geq 2$, then by choosing

$$x_0 = P[\underbrace{0 \ldots \ldots 0}_{n_1 + \ldots + n_{j-1}} \quad 0 \quad \delta \quad 0 \ldots 0]^T,$$

we have, from (6.15),

$$|x_k| = |A^k x_0| = |PJ^k P^{-1} x_0|$$

$$= |P[\underbrace{0 \ldots \ldots 0}_{n_1 + \ldots + n_{j-1}} \quad k\lambda_j^{k-1}\delta \quad \lambda_j^k\delta \quad 0 \ldots 0]^T|$$

$$= \delta\left[|k\lambda_j^{k-1}|^2\left(\sum_{r=1}^{n} P_{r,n_1 + \ldots + n_{j-1}+1}^2\right)\right.$$

$$\left. + |\lambda_j^k|^2\left(\sum_{r=1}^{n} p_{r,n_1 + \ldots + n_{j-1}+2}^2\right)\right]^{1/2}$$

$$> k\delta\left(\sum_{r=1}^{n} p_{r,n_1 + \ldots + n_{j-1}+1}^2\right)^{1/2} \to \infty,$$

as $k \to \infty$ for each $\delta > 0$, where $P = [p_{rs}]$, because the $(n_1 + \ldots + n_{j-1} + 1)$st column of P cannot be identically zero, P being nonsingular. Since $\delta > 0$ is arbitrary, the system is not stable in the sense of Lyapunov about 0. This completes the proof of the theorem.

Remark 6.6 If the system (6.12) is asymptotically stable about 0, we have actually proved that $|x_k|$ decays to zero exponentially fast. There is another way to see this behavior. Consider A as a transformation from $\mathbb{R}^n$ into $\mathbb{R}^n$. Then we may consider the *operator norm* of this transformation defined by

$$\|A\| = \sup\{|Ax| : |x| = 1\}$$

(which really means the maximum of the lengths of the vectors Ax among all unit vectors x in $\mathbb{R}^n$). There is an important result that relates $\|A^k\|$ to the magnitudes of the eigenvalues of A. If λ_js are the eigenvalues of A, this result, called the *Spectral Radius Theorem*, says that the sequence $\{\|A^k\|^{1/k}\}$ converges as $k \to \infty$, and

$$\lim_{k \to \infty} \|A^k\|^{1/k} = \max|\lambda_j|.$$

Hence, if all $|\lambda_j| < 1$, then for any ρ with $|\lambda_j| < \rho < 1$, we have

$$\|A^k\| \le \rho^k$$

for all large values of k, so that (Exercise 6.13)

$$|x_k| = |A^k x_0| \le \|A^k\|\, |x_0| \le |x_0|\rho^k \ . \tag{6.16}$$

Inequality (6.16) is analogous to inequality (6.7) for continuous-time systems. It is, therefore, very natural to consider discrete-time time-varying free linear systems and to characterize the ones that are "exponentially stable" about 0 (i.e., satisfying (6.16)). We leave this as an exercise to the reader (Exercise 6.15).

6.4 Input-Output Stability of Continuous-Time Linear Systems

We next consider *input-output stability* of a non-free linear system. It will be interesting to see that although there is a very tight relationship between asymptotic state-stability (i.e. asymptotic stability of a free system) and the input-output stability that we are going to discuss, there does exist an input-output stable linear system that is *not* state-stable, as mentioned in Sect. 5.4. The main reason is a pole-zero cancellation (Theorem 5.2 and the example following Theorem 6.8).

We will first consider the continuous-time state-space description. If we have an input function $u(t)$ which is bounded for all $t \ge t_0$, one would certainly hope to have a bounded output response $v(t)$. This is essentially the definition of input-output stability (or bounded-input bounded-output stability). Recall that the output v not only depends on the state vector x, but sometimes also depends on the input u *directly*, as described by the transfer matrix $D(t)$ in (1.7). Since u is supposed to be bounded and an unbounded transfer matrix is unlikely and very undesirable, the term $D(t)u$ is usually discarded in the discussion of input-output stability. That is, we will consider the state-space description

$$\begin{aligned}\dot{x} &= A(t)x + B(t)u \\ v &= C(t)x \ .\end{aligned} \tag{6.17}$$

Definition 6.6   A linear system with the state-space description (6.17) is said to be *input-output stable* about an equilibrium point x_e (or $I - O$ stable, for short), if for any given positive constant M_1, there exists a positive constant M_2, such that whenever $x(t_0) = x_e$ and $|u(t)| \le M_1$ for all $t \ge t_0$, we have $|v(t)| \le M_2$ for all $t \ge t_0$.

In view of Remark 6.1, we will always assume the equilibrium point x_e to be 0. Hence, the input-output relation can be expressed with the aid of the transition matrix by

$$v(t) = \int_{t_0}^{t} C(t)\Phi(t, s)\, B(s)u(s)\, ds \ , \tag{6.18}$$

see (2.4). This relationship describes the I−O stability completely. For convenience, we introduce the notation

$$h^*(t, s) = C(t)\Phi(t, s)B(s) \tag{6.19}$$

so that (6.18) becomes

$$v(t) = \int_{t_0}^{t} h^*(t, s)u(s)\,ds \ . \tag{6.20}$$

Theorem 6.5 *A linear system described by (6.17) is I−O stable if and only if there exists a positive constant $M(t_0)$ such that $h^*(t, s)$ satisfies*

$$\int_{t_0}^{t} |h^*(t, s)|\,ds \le M(t_0) \tag{6.21}$$

for all $t \ge t_0$.

One direction is clear. If $|u(t)| \le M_1$ for all $t \ge t_0$ and (6.21) is satisfied, then by using the inequality in Exercise 6.6 and Schwarz's inequality, we have, from (6.20),

$$|v(t)| \le \int_{t_0}^{t} |h^*(t, s)u(s)|\,ds$$

$$\le \int_{t_0}^{t} |h^*(t, s)|\,|u(s)|\,ds$$

$$\le M_1 \int_{t_0}^{t_1} |h^*(t, s)|\,ds \le M_1 M(t_0) \ .$$

To prove the converse, we assume, on the contrary, that (6.21) is not satisfied but $|u(t)| \le M_1$ implies $|v(t)| \le M_2$ for all $t \ge t_0$. Let $h_{ij}(t, s)$ be the (i, j)th entry of the $q \times p$ matrix $h^*(t, x)$. Since (6.21) is not satisfied for each (arbitrarily large) positive constant N we can choose $t_1 > t_0$ such that

$$\int_{t_0}^{t_1} |h^*(t_1, s)|\,ds > pqN \ .$$

Hence, we have

$$pqN < \int_{t_0}^{t_1} \left[\sum_{j=1}^{p} \sum_{i=1}^{q} |h_{ij}(t_1, s)|^2 \right]^{1/2} ds$$

$$\le \int_{t_0}^{t_1} \sum_{j=1}^{p} \sum_{i=1}^{q} |h_{ij}(t_1, s)|\,ds$$

$$\le pq \int_{t_0}^{t_1} |h_{\alpha\beta}(t_1, s)|\,ds$$

which implies

$$\int_{t_0}^{t_1} |h_{\alpha\beta}(t_1, s)| \, ds > N \tag{6.22}$$

for some (α, β), where $1 \leq \alpha \leq q$ and $1 \leq \beta \leq p$. Now choose $u = [0 \ldots 0 \text{ sgn } h_{\alpha\beta}(t_1, s) \, 0 \ldots 0]^T$, where sgn $h_{\alpha\beta}(t_1, s)$ is placed at the βth component of u and denotes the function which is 1 if $h_{\alpha\beta}(t_1, s)$ is positive, 0 if $h_{\alpha\beta}(t_1, s)$ is 0, and -1 if $h_{\alpha\beta}(t_1, s)$ is negative (usually called the *signum function*). Then (6.20 and 22) give

$$|v(t_1)|^2 = \left| \int_{t_0}^{t_1} h^*(t_1, s) u(s) \, ds \right|^2$$

$$= \left[\int_{t_0}^{t_1} |h_{\alpha\beta}(t_1, s)| \, ds \right]^2 + \sum_{i \neq a} \left[\int_{t_0}^{t_1} h_{i\beta}(t_1, s) \text{ sgn } h_{\alpha\beta}(t_1, s) \, ds \right]^2$$

$$\geq \left[\int_{t_0}^{t_1} |h_{\alpha\beta}(t_1, s)| \, ds \right]^2 > N^2 \ .$$

That N was arbitrarily chosen contradicts the assumption $|v(t)| \leq M_2$ for all $t \geq t_0$. This completes the proof of the theorem.

Since the $q \times p$ matrix $h^*(t, s)$ defined in (6.19) plays a very important role in characterizing I $-$ O stability, it is worth investigating this function in the time-invariant setting.

Let A, B, C in (6.17) be constant $n \times n$, $n \times p$, and $q \times n$ matrices. Then (6.19) becomes

$$h^*(t, s) = Ce^{(t-s)A}B \ .$$

Note that the right-hand side can be considered as a function of one variable $(t-s)$. Hence, we can introduce the $q \times p$ matrix-valued function

$$h(t) = Ce^{tA}B \ , \tag{6.23}$$

so that $h^*(t, s) = h(t-s)$. For convenience, we consider $t_0 \geq 0$ and for any input $u(t)$, we define $u(t)$ to be 0 for $t < t_0$. Then (6.20) can be written as

$$v(t) = \int_{t_0}^{t} h(t-s) u(s) \, ds = \int_{0}^{t} h(t-s) \, u(s) \, ds \tag{6.24}$$

$$= (h * u)(t) \ ,$$

called the *convolution* of h with u. Since the Laplace transform of a convolution is the product of the Laplace transforms, we can conclude that

$$(\mathcal{L}h)(s) = H(s) = \frac{C(sI - A)^* B}{\det(sI - A)} \tag{6.25}$$

is the *transfer function* of the system [cf. (5.11)]; or equivalently, $h(t)$ in (6.23) is the inverse Laplace transform of the transform function $H(s)$. That is, $h(t)$ is the *impulse response* of the time-invariant linear system (6.17).

Theorem 6.6   *The impulse response $h(t)$ satisfies*

$$\int_{t_0}^{t} |h(t-s)|\, ds = \int_{0}^{t-t_0} |h(\tau)|\, d\tau \leq M(t_0) < \infty$$

for all $t \geq t_0$ if and only if all the poles of the transfer function $H(s)$ lie on the left (open) half s-complex plane.

In view of Theorem 6.5, an equivalent statement of the above theorem is the following.

Theorem 6.7   *A time-invariant linear system described by (6.17) is $I - O$ stable if and only if all the poles of its transfer function lie on the left (open) half complex plane.*

It is sufficient to prove Theorem 6.6. Imitating the argument that yields (6.5), we have

$$h(t) = \sum_{j=1}^{d} \sum_{l=0}^{n_j - 1} \frac{t^l}{l!} e^{\lambda_j t} Q_{lj} \tag{6.26}$$

where $\lambda_1, \ldots, \lambda_d$ are the poles of $H(s)$ with multiplicities $n_1, \ldots, n_d$ respectively, and Q_{lj} are constant $q \times p$ matrices (Exercise 6.16). The theorem then follows from standard estimates (Exercise 6.17).

Note that the poles of the transfer function $H(s)$ are eigenvalues of A, but since there is a possibility of pole-zero cancellation of $H(s)$ in the expression (6.25), the converse does not hold. However, if the linear system is both completely controllable and observable, Theorem 5.2 tells us that the set of poles of $H(s)$ is the same as the collection of eigenvalues of A. Hence, as a consequence of Theorems 6.2 and 7, we immediately have the following result.

Theorem 6.8   *Let the time-invariant system described by (6.17) be completely controllable and observable. Then the system is $I - O$ stable if and only if the free linear system $\dot{x} = Ax$ is asymptotically stable about the equilibrium point 0.*

Let us return to the example (5.12) considered in Sect. 5.4; namely,

$$A = \begin{bmatrix} -2 & 1 \\ 3 & 0 \end{bmatrix}, \quad B = \begin{bmatrix} 1 \\ -1 \end{bmatrix}, \quad C = [0 \quad -1].$$

Recall that the eigenvalues of A are 1 and -3 so that the free linear system is *not* (*state-*) *stable*, but the only pole of $H(s)$ is -3 so that it is *input-output stable*. Indeed, this system is observable but is *not* controllable. In addition the transition matrix is

$$e^{tA} = \frac{1}{4} \begin{bmatrix} 3e^{-3t} + e^t & -e^{-3t} + e^t \\ -3e^{-3t} + 3e^t & e^{-3t} + 3e^t \end{bmatrix}$$

(which is unbounded), but the impulse response

$$h(t) = Ce^{tA}B = e^{-3t}$$

certainly satisfies

$$\int_0^t |h(t-s)|\, ds < \tfrac{1}{3} \ .$$

for all $t \geq 0$.

6.5 Input-Output Stability of Discrete-Time Linear Systems

We next consider discrete-time linear systems. Only time-invariant settings will be discussed (cf. Exercise 6.18 for time-varying systems). That is, we now study the state-space description

$$\begin{aligned} x_{k+1} &= Ax_k + Bu_k \\ v_k &= Cx_k \end{aligned} \tag{6.27}$$

As before, we have assumed the transfer matrix D to be 0.

Definition 6.7 A linear system with the state-space description (6.27) is *input-output stable* about 0 (or *I − O stable*, for short), if there exists a positive constant M such that whenever $x_0 = 0$ and $|u_k| \leq 1$ for $k = 0, 1, \ldots$, we have $|v_k| \leq M$ for $k = 0, 1, \ldots$.

Since $x_0 = 0$, we have the input-output relationship

$$v_k = \sum_{l=0}^{k-1} h_{k-l} u_l \tag{6.28}$$

where the $q \times p$ matrices h_j are defined by

$$h_j = CA^{j-1}B, \quad j = 1, 2, \ldots, \tag{6.29}$$

which we will call the *impulse response* sequence of the system. Analogous to Theorem 6.5, we have the following test for $I-O$ stability (Exercise 6.19).

Theorem 6.9   *A discrete-time time-invariant system described by* (6.27) *is $I-O$ stable if and only if there exists a positive constant K such that*

$$\sum_{j=1}^{k} |h_j| \le K$$

for all $k = 1, 2, \ldots$.

The input-output relationship (6.28) can be thought of as the convolution of the sequence of $q \times p$ matrices $\{h_j\}$ and the sequence of p-vectors $\{u_j\}$. In fact, if we define

$$h_j = 0, \quad u_l = 0 \ ,$$

for $j \le 0$ and $l < 0$, then (6.28) can be written as

$$v_k = \sum_{l=-\infty}^{\infty} h_{k-l} u_l \ .$$

Now, taking the z-transforms of both sides yields:

$$V(z) = Z\{v_k\} = \sum_{k=0}^{\infty} v_k z^{-k}$$

$$= \sum_{k=0}^{\infty} \left(\sum_{l=-\infty}^{\infty} h_{k-l} u_l \right) z^{-k}$$

$$= \sum_{l=-\infty}^{\infty} \sum_{j=-l}^{\infty} h_j u_l z^{-k-l}$$

$$= \sum_{j=1}^{\infty} h_j z^{-j} \sum_{l=0}^{\infty} u_l z^{-l} = H(z) U(z)$$

where

$$H(z) = \sum_{j=1}^{\infty} h_j z^{-j} \tag{6.30}$$

is the *transfer function* of the discrete-time system. We have already mentioned in Sect. 5.3 that the z-transform properties are completely analogous to the Laplace transform properties; hence $H(z)$ has exactly the same formulation as (5.11); that is

$$H(z) = \frac{C(zI - A)^* B}{\det(zI - A)} \tag{6.31}$$

(Exercise 6.20). Write the $q \times p$ matrix h_j as

$$h_j = [c_{lm}^{(j)}]$$

$1 \leq l \leq q$, $1 \leq m \leq p$,   and $j = 1, 2, \ldots$, so that

$$H(z) = \left[\sum_{j=1}^{\infty} c_{lm}^{(j)} z^{-j} \right]. \tag{6.32}$$

It is obvious that

$$\sum_{j=1}^{\infty} |h_j| < \infty$$

if and only if

$$\sum_{j=1}^{\infty} |c_{lm}^{(j)}| < \infty \tag{6.33}$$

for all $1 \leq l \leq q$ and $1 \leq m \leq p$. Also, since each power series

$$\sum_{j=1}^{\infty} c_{lm}^{(j)} z^{-j} \tag{6.34}$$

is a rational function in z^{-1} from (6.31, 32), the inequality (6.33) is satisfied if and only if the power series (6.34) is an analytic function in (a neighborhood of) $|z^{-1}| \leq 1$ or $|z| \geq 1$, $1 \leq l \leq q$, $1 \leq m \leq p$, or equivalently, all poles of $H(z)$ in (6.31) lie in the open unit disk $|z| < 1$ (Exercise 6.21 where $w = z^{-1}$). An application of Theorem 6.9 yields the following result.

Theorem 6.10   *A discrete-time time-invariant system described by (6.27) is $I - O$ stable if and only if all the poles of its transfer function $H(z)$ lie in $|z| < 1$.*

Again, if there is no pole-zero cancellation in (6.31), then the set of poles of $H(z)$ coincides with the collection of eigenvalues of A. Hence, Theorems 5.2 and 6.4 together yield the following result.

Theorem 6.11   *Let the discrete-time time-invariant system described by (6.27) be completely controllable and observable. Then it is $I - O$ stable if and only if the free linear system $x_{k+1} = Ax_k$ is asymptotically stable about 0.*

Note that if a discrete-time free linear system is asymptotically stable about 0, then the corresponding state-space description is $I - O$ stable. However, without the additional assumption on both complete controllability and observability, the converse usually does not hold (Exercises 6.22, 23).

Exercises

6.1   Determine all equilibrium points of the free linear system with system matrix:

$$\text{(a) } A = \begin{bmatrix} 0 & 0 & 1 \\ 0 & 0 & 0 \\ 0 & 0 & 0 \end{bmatrix}, \qquad \text{(b) } A = \begin{bmatrix} 0 & 0 & 0 \\ 0 & 0 & 0 \\ 0 & 0 & 1 \end{bmatrix}.$$

6.2   Determine all equilibrium points of the time-varying free linear system with system matrix:

$$\text{(a) } A(t) = \begin{bmatrix} 0 & t \\ 0 & 0 \end{bmatrix}, \qquad \text{(b) } A(t) = \begin{bmatrix} t - t_0 & 1 \\ 0 & 0 \end{bmatrix}.$$

6.3   If $A(t)$ is nonsingular for some $t > t_0$, show that the only equilibrium point of $\dot{x} = A(t)x$ is 0.

6.4   Let E and F be $m \times n$ and $n \times p$ matrices. Prove the following Schwarz's inequality: $|EF|_2 \le |E|_2 |F|_2$. Compare with Exercise 2.8.

6.5   Use the triangle inequality in Exercise 2.8 to show:

$$|\,|A|_p - |B|_p| \le |A + B|_p\ ,$$

where A and B are matrices of the same order and $p \ge 1$.

6.6   Let $F(t)$ be an $m \times n$ matrix-valued continuous function of t. Show that

$$\left| \int_a^b F(t)\, dt \right|_p \le \int_a^b |F(t)|_p\, dt\ .$$

(*Hint*: Use Riemann sums and Exercise 2.8).

6.7   If a free linear system is asymptotically stable about 0, show that (6.3) must be satisfied. (This completes the proof of Theorem 6.1).

6.8   Let a and b be positive constants. Prove:

(a) $\lim\limits_{t \to +\infty} e^{-at} t^b = 0$ and

(b) $\lim\limits_{m \to \infty} m^a c^m = 0$ if $|c| < 1$.

6.9   Show that if $|f(t)| \le M \exp[-at]\ t^b$ for all $t \ge 0$ and $0 < c < a$, then $|f(t)| \le \exp(-ct)$ for all large values of t.

6.10   Prove Theorem 6.2 by using Exercise 6.8 and Theorem 6.1.

6.11   Consider the Jordan canonical forms:

$$J_1 = \begin{bmatrix} \lambda & & & \\ & \ddots & & \\ & & \ddots & \\ & & & \lambda \end{bmatrix} \quad \text{and} \quad J_2 = \begin{bmatrix} \lambda & 1 & & \\ & \ddots & \ddots & \\ & & \ddots & 1 \\ & & & \lambda \end{bmatrix}$$

where the unspecified entries are 0. Determine J_1^k and J_2^k and show that $\lim_{k \to \infty} |J_1^k|_2 = \lim_{k \to \infty} |J_2^k|_2 = 0$ if $|\lambda| < 1$; and $|J_1^k|_2$ is bounded but $|J_2^k|_2$ is not if $|\lambda| = 1$.

6.12 Let

$$A = \begin{bmatrix} 0 & -1 \\ 1 & 0 \end{bmatrix} .$$

Discuss the stability (in the sense of Lyapunov) about 0 of the free linear systems:
(a) $\dot{x} = Ax$ and (b) $x_{k+1} = Ax$.

6.13 Let $\|A\|$ be the operator norm of the matrix A. Show the following:
(a) $\|A\| \le |A|_2$
(b) If λ is an eigenvalue of A, then $|\lambda| \le \|A\|$.
(c) $\|A + B\| \le \|A\| + \|B\|$ and $\|\alpha A\| = |\alpha| \, \|A\|$.

6.14 Let A be an $n \times n$ constant matrix. Show that $x_{k+1} = Ax_k$ is stable about 0 if and only if $\|A^k\|$ is bounded for all k, and is asymptotically stable about 0 if and only if $\|A^k\| \to 0$ as $k \to \infty$.

6.15 Define asymptotic and exponential stability for discrete-time time-varying free linear systems. Give criteria for testing these stabilities.

6.16 Derive (6.26) by using partial fractions.

6.17 Prove Theorem 6.6 by following the proof of Theorem 6.2. Note, however, that since we require a uniform bound on the integral, even simple eigenvalues with zero real part are not permissible.

6.18 Discuss $I - O$ stability for discrete-time time-varying linear systems and formulate an analog of Theorem 6.5.

6.19 Prove Theorem 6.9 by imitating the proof of Theorem 6.5.

6.20 Following the derivation of (5.11), derive (6.31).

6.21 Let $f(w) = \sum_0^\infty a_n w^n$ be a rational function which is analytic at $w = 0$. Prove that the radius of convergence of the power series is larger than 1 if and only if $\sum_0^\infty |a_n| < \infty$.

6.22 Give an example of a completely controllable $I - O$ stable time-invariant linear system which is not asymptotically state-stable (i.e. with a corresponding asymptotically unstable free linear system).

6.23 Give an example of an observable $I - O$ stable time-invariant linear system which is not asymptotically state-stable.

7. Optimal Control Problems and Variational Methods

In the previous discussions on controllability, we have been concerned with the possibility of bringing a state (vector) from an initial position to an assigned position, namely the target, in a finite amount of time. In practice, many factors must be brought into consideration. For instance, the state may not be allowed to travel outside a certain region and the control (function) has certain limited capacity. Another important consideration is that there are certain quantities that we wish to optimize. Usually the quantities to be minimized are time, fuel, energy, cost, etc. and those to be maximized include speed, efficiency, profit, etc. The problem under consideration is, therefore, to optimize a quantity, called a *functional*, which usually depends on the control function, the state vector, and the time parameter, and at the same time, to satisfy certain constraints, namely: the control equation of the state-space description, a region the state vector is confined to, and an admissible collection of functions to which the control function belongs.

7.1 The Lagrange, Bolza, and Mayer Problems

Let us consider the continuous-time models. As usual, J denotes the time interval, $x = x(t)$ an n-dimensional state vector, and $u = u(t)$ a p-dimensional vector-valued control function; but instead of the linear control equation of the state-space description, let us consider the more general control equation:

$$\dot{x} = f(x, u, t) \tag{7.1}$$

where f is a vector-valued (linear or nonlinear) function defined on $\Omega \times J$, with $\Omega \subset \mathbb{R}^n$ and $J = [t_0, \infty)$. Let $x(t)$ be confined to a set $X \subset \mathbb{R}^n$ for all $t \in J$ and let U be a collection of vector-valued functions containing $u = [u_1 \ldots u_p]^T$. Typically, we might have:

$$a_i \leq u_i(t) \leq b_i, \quad i = 1, \ldots, p \quad \text{and} \quad t \in J \quad \text{or}$$

$$|u(t)|_2 \leq c, \quad t \in J ,$$

etc. Let us study the optimization of the functional

$$F(u) = \int_{t_0}^{t_1} g(x, u, t) \, dt \tag{7.2}$$

where $g(x, u, t)$ is a scalar-valued continuous function defined on $X \times U \times J$ (i.e. $x \in X$, $u \in U$, and $t \in J$), and x depends on u according to (7.1). This is usually called the *Lagrange problem*. If g does not explicitly depend on t, then the domain of g is simply reduced to $X \times U$, and if g depends only on u directly, its domain of definition is further reduced to U, etc. Examples of this optimal control problem are:

i) minimum-energy control problem, with

$$g(u) = u^T R(t) u \ ,$$

where $R(t)$ is a symmetric and non-negative definite matrix;
ii) minimum-fuel control problem, with

$$g(u) = |u|_1 \ ;$$

iii) minimum-time control problem, with

$$g(u) = 1$$

(where t_1 depends on u).

The functional $F(u)$ in (7.2) to be optimized (minimized or maximized) is called a *cost functional* (or *penalty functional*). Since $\min \{F(u)\} = -\max \{-F(u)\}$, there is no distinction between the two optimization processes. For this reason, we will usually discuss the minimization problem. If we add another term to (7.2), say, by considering the functional

$$F(u) = h(t_1, x(t_1)) + \int_{t_0}^{t_1} g(x, u, t) \, dt \ ,$$

we have what is usually called the *Bolza problem*. By considering the functional

$$F(u) = h(t_1, x(t_1))$$

alone, we have what is called the *Mayer problem*. Of course, in all the above statements, we must treat the indicated variables t_1, $x(t_1)$, and x as functions of the control function u which is restricted to U, and remember that x satisfies (7.1) with the initial condition $x(t_0) = x_0$ such that $x \in X$. It is clear that the Lagrange and Mayer problems are special cases of the Bolza problem. On the other hand, by introducing an extra state variable, it can be shown that the Bolza problem can be changed to the Lagrange problem or the Mayer problem (Exercise 7.2).

It is also interesting to mention that the three problems mentioned here are special cases of the so-called *Pontryagin function*:

$$F(u) = c^T x(t_1) \ , \tag{7.3}$$

where $c^T = [c_1 \ldots c_n]$ is a constant row vector. For the Lagrange problem, for instance, we may introduce a state variable x_{n+1} defined by

$$x_{n+1}(t) = \int_{t_0}^{t} g(x, u, \tau) d\tau$$

and consider the new state vector

$$y = \begin{bmatrix} x \\ x_{n+1} \end{bmatrix}$$

in R^{n+1}, so that with $c^T = [0 \ldots 0 \ 1]$, we have

$$c^T y(t_1) = x_{n+1}(t_1) = \int_{t_0}^{t_1} g(x, u, \tau) d\tau \ .$$

Of course, the new state vector must satisfy the control equation:

$$\dot{y} = \begin{bmatrix} \dot{x} \\ \dot{x}_{n+1} \end{bmatrix} = \begin{bmatrix} f(x, u, t) \\ g(x, u, t) \end{bmatrix} := \tilde{f}(y, u, t) \ .$$

If the terminal time t_1 is free and the terminal state $x(t_1)$ is restricted, then both these quantities depend on the control function u, and the optimal control problem is, in general, very difficult to solve. In this chapter we do not intend to solve the most general problem, but rather consider the special case where t_1 is fixed and no restriction is imposed on $x(t_1)$. The more general problems will be studied in the next three chapters.

7.2 A Variational Method for Continuous-Time Systems

More precisely, the problem we will study here is to find necessary conditions that the *optimal control function* u^* and its corresponding *optimal trajectory* (or *state*) x^* defined by

$$\begin{cases} F(u^*) = \min\{F(u): u \in U\} \ , \\ \dot{x}^* = f(x^*, u^*, t), \quad t_0 \le t \le t_1 \ , \\ x^*(t_0) = x_0 \end{cases} \tag{7.4}$$

must satisfy, where $F(u)$ is defined by (7.2) with initial condition $x(t_0) = x_0$ and fixed terminal time t_1 such that $\dot{x} = f(x, u, t)$ for $t_0 \le t \le t_1$.

A classical approach to this problem is via the calculus of variations. This method, however, has its limitations. Since partial derivatives must be taken, we require the functions $f(x, u, t)$ and $g(x, u, t)$ in (7.2) to be continuous and have continuous partial derivatives with respect to all components of x and u. In addition, we require that the admissible set U of control functions is "complete" in the space of vector-valued continuous functions $k(t) \in R^p$, $t \in J$, in the sense that whenever

$$\int_{t_0}^{t_1} k^T(t)\, \eta(t)\, dt = 0$$

for all $\eta \in U$, then we must have $k = 0$. An example of such a set U is the collection of all vector-valued piecewise continuous functions u with $|u| < 1$ (Exercise 7.3). Since we will be taking the "variations" with respect to functions in U, it is also convenient to assume that every function u in U is *interior* to U, in the sense that for each $\eta \in U$, there exists an $\varepsilon_0 > 0$ such that $(u + \varepsilon \eta) \in U$ for all $|\varepsilon| < \varepsilon_0$. Hence, if $l(u, t)$ is a vector- or scalar-valued function, with continuous first partial derivatives with respect to the components of u, say $l(u, t) = [l_1 \ldots l_m]^T$ and $\eta \in U$, then the *variation* of $l = l(u, t)$ with respect to u along η is defined by

$$\delta l = \delta_\eta l = \lim_{\varepsilon \to 0} \frac{1}{\varepsilon} \left[l(u + \varepsilon \eta,\ t) - l(u,\ t) \right] \tag{7.5}$$

$$= \frac{\partial l}{\partial u} \eta\ ,$$

where, using the notation $u = [u_1 \ldots u_p]^T$, the $m \times p$ matrix $\partial l / \partial u$ is given by

$$\frac{\partial l}{\partial u} = \begin{bmatrix} \dfrac{\partial l_1}{\partial u_1} & \cdots & \dfrac{\partial l_1}{\partial u_p} \\ \vdots & & \vdots \\ \dfrac{\partial l_m}{\partial u_1} & \cdots & \dfrac{\partial l_m}{\partial u_p} \end{bmatrix} \tag{7.6}$$

In particular, if $l = l$ is a scalar-valued function, then $\partial l / \partial u$ is a row-vector which is usually called the *gradient* of l with respect to u. We will take the variations of both the control equation (7.1) and the cost functional (7.2). Let us use the notation

$$\xi = \delta x\ .$$

Then from (7.1) the variation of $\dot{x}$ becomes (Exercise 7.4):

$$\dot{\xi} = \frac{\partial f}{\partial x} \xi + \frac{\partial f}{\partial u} \eta\ . \tag{7.7}$$

This equation can be "solved" by using the state transition equation (2.4). Since the initial state $x(t_0) = x_0$ is unchanged as long as the control functions are chosen from U, we have $\xi(t_0)=0$ from the definition (7.5). Hence, if $\Phi(t, s)$ denotes the transition matrix of (7.7), we have

$$\xi(t)= \int_{t_0}^{t} \Phi(t, \tau)\frac{\partial f}{\partial u}(x, u, \tau)\, \eta(\tau)\, d\tau \ . \tag{7.8}$$

On the other hand, taking the variation of the cost functional (7.2) with respect to u along η, and keeping in mind that we have assumed a fixed final time t_1, we have

$$\delta_\eta F(u)= \int_{t_0}^{t_1}\left[\frac{\partial g}{\partial x}(x, u, t)\,\xi(t)+ \frac{\partial g}{\partial u}(x, u, t)\,\eta(t)\right]dt \ . \tag{7.9}$$

To minimize the cost functional $F(u)$, it is necessary that $\delta_\eta F(u)=0$ for all η in U. Hence, putting (7.8) into (7.9), interchanging the integrals, and using the completeness of U in the space of continuous functions, we arrive at the following necessary condition for an optimal $F(u)$ (Exercise 7.5):

$$\frac{\partial g}{\partial u}(x^*, u^*, \tau)+ \int_{\tau}^{t_1} \frac{\partial g}{\partial x}(x^*, u^*, t)\Phi(t, \tau)\frac{\partial f}{\partial u}(x^*, u^*, \tau)\, dt =0 \ . \tag{7.10}$$

Here, $t_0\le\tau\le t_1$, and u^* and x^* denote an optimal control function and its corresponding optimal trajectory (state).

In order to be able to work with the equation (7.10), we introduce an n-dimensional vector-valued function $p=p(t)$, called a *costate* which is defined, for any pair (u, x) satisfying (7.1), to be the unique solution of the initial value problem

$$\begin{cases} \dot{p}= -\left[\frac{\partial f}{\partial x}(x, u, \tau)\right]^{T} p -\left[\frac{\partial g}{\partial x}(x, u, \tau)\right]^{T} \\ p(t_1)=0 \ . \end{cases} \tag{7.11}$$

Let p^* be the costate corresponding to the optimal pair (u^*, x^*) and call it an *optimal costate*. We also call (7.11) the *costate equation*. Let $\Psi(\tau, t)$ be its transition matrix. By Lemma 4.1, we have $\Psi(\tau, t)=\Phi^{T}(t, \tau)$ where $\Phi(t, \tau)$ is the transition matrix of (7.7). Hence, we have

$$p(\tau)= - \int_{t_1}^{\tau} \Phi^{T}(t, \tau)\left[\frac{\partial g}{\partial x}(x, u, t)\right]^{T} dt$$

so that (7.10) becomes

$$\frac{\partial g}{\partial u}(x^*, u^*, t) + p^{*T}(t)\frac{\partial f}{\partial u}(x^*, u^*, t) = 0, \quad t_0 \le t \le t_1 \ .$$

That is, if we define the functional

$$H(x, u, p, t) = g(x, u, t) + p^T f(x, u, t) \tag{7.12}$$

which is called the *Hamiltonian*, a quantity that often occurs in classical mechanics, then a necessary condition for u^* and x^* to be optimal is that

$$\frac{\partial H}{\partial u}(x^*, u^*, p^*, t) = 0, \quad t \in [t_0, t_1] \ . \tag{7.13}$$

Let us restate this result.

Theorem 7.1 *A necessary condition for the pair (u^*, x^*) to satisfy*

$$\begin{cases} F(u^*) = \min[F(u): u \in U] \ , \\ \dot{x}^* = f(x^*, u^*, t), \quad t_0 \le t \le t_1 \ , \\ x^*(t_0) = x_0 \end{cases} \tag{7.14}$$

where $F(u)$ is given by (7.2) with initial condition $x(t_0) = x_0$ and fixed terminal time such that (u, x) satisfies (7.1) is the existence of a costate p such that the corresponding Hamiltonian defined by (7.12) satisfies (7.13).

Note that if g is independent of x, then since (7.11) has a unique solution, the costate p is always zero, so that we have the following result.

Corollary 7.1 *A necessary condition for the pair (u^*, x^*) to satisfy (7.14) where*

$$F(u) = \int_{t_0}^{t_1} g(u, t)\, dt$$

such that $\dot{x} = f(x, u, t)$, $x(t_0) = x_0$ and t_1 being fixed is that $\partial g(u^*, t)/\partial u = 0$ for $t_0 \le t \le t_1$.

Hence, if g does not depend on the state, as in the case of the minimum-energy control problem, and the terminal time and state are fixed, determining (u^*, x^*) is usually fairly easy. However, in many problems in control theory, the cost functional depends on the state vector x. Let E, $Q(t)$ and $R(t)$ be symmetric and nonnegative definite matrices of appropriate dimensions. The so-called *linear regulator* problem (with a linear state-space description) involves a cost functional of the form

$$F(u) = \tfrac{1}{2} x^T(t_1) E x(t_1) + \tfrac{1}{2} \int_{t_0}^{t_1} [x^T(t)Q(t)x(t) + u^T(t)R(t)u(t)]\, dt \ ; \tag{7.15}$$

and the *linear servomechanism* (again with a linear state-space description) is a problem of approximating a certain desired trajectory $y = y(t)$ by minimizing the cost functional

$$F(u) = \tfrac{1}{2} \int_{t_0}^{t_1} \{ [y(t) - x(t)]^T Q(t) [y(t) - x(t)] + u^T(t) R(t) u(t) \} \, dt \qquad (7.16)$$

(Exercises 7.8 and 9).

7.3 Two Examples

To illustrate the method described in Theorem 7.1, let us consider the one-dimensional control equation (of a state-space description)

$$\dot{x} = x + u \; ,$$

with the initial state $x(0) = 1$, and determine the optimal control function u^* and its corresponding trajectory x^* when the cost functional to be minimized is

$$F(u) = \tfrac{1}{2} \int_0^1 [x^2(t) + u^2(t)] \, dt \; .$$

The costate equation is clearly

$$\dot{p} = -p - x$$

$$p(1) = 0$$

since $\partial g / \partial x = x$. Therefore, combining this with the original control equation, we have a so-called "two-point boundary value problem":

$$\begin{bmatrix} \dot{x} \\ \dot{p} \end{bmatrix} = \begin{bmatrix} 1 & 0 \\ -1 & -1 \end{bmatrix} \begin{bmatrix} x \\ p \end{bmatrix} + \begin{bmatrix} 1 \\ 0 \end{bmatrix} u$$

$$x(0) = 1, \quad p(1) = 0 \; .$$

Since the Hamiltonian is

$$H(x, u, p, t) = \tfrac{1}{2}(x^2 + u^2) + p(x + u)$$

and $\partial H / \partial u = u + p$, we also have, for optimality,

$$p^* = -u^* \; .$$

That is, we must solve the two-point boundary value problem:

$$\begin{bmatrix} \dot{x}^* \\ \dot{p}^* \end{bmatrix} = \begin{bmatrix} 1 & -1 \\ -1 & -1 \end{bmatrix} \begin{bmatrix} x^* \\ p^* \end{bmatrix}$$

$$x^*(0) = 1, \quad p^*(1) = 0 \; .$$

An elementary calculation shows that

$$u^*(t) = -p^*(t) = \frac{\sqrt{2}-1}{(3-2\sqrt{2})e^{2\sqrt{2}}+1}e^{\sqrt{2}t} - \frac{\sqrt{2}+1}{(3+2\sqrt{2})e^{-2\sqrt{2}}+1}e^{-\sqrt{2}t}$$

and

$$x^*(t) = \frac{1}{(3-2\sqrt{2})e^{2\sqrt{2}}+1}e^{\sqrt{2}t} + \frac{1}{(3+2\sqrt{2})e^{-2\sqrt{2}}+1}e^{-\sqrt{2}t}$$

provided, of course, that $u^* \in U$ (Exercise 7.6).

However, a two-dimensional problem is much more complicated. For instance, consider the initial valued control problem

$$\dot{x} = \begin{bmatrix} 0 & 1 \\ 0 & 0 \end{bmatrix} x + \begin{bmatrix} 0 \\ 1 \end{bmatrix} u$$

$$x(0) = [1 \quad 0]^T$$

with cost functional

$$F(u) = \frac{1}{2}\int_0^1 [x^T(t)x(t) + u^2(t)] \, dt$$

to be minimized. The costate equation here is

$$\begin{cases} \dot{p} = \begin{bmatrix} 0 & 0 \\ -1 & 0 \end{bmatrix} p - x \\ p(1) = 0 \end{cases}$$

and

$$\frac{\partial H}{\partial u}(x^*, u^*, p^*, t) = u^* + p^{*T}\begin{bmatrix} 0 \\ 1 \end{bmatrix} = 0 \ .$$

Hence, we must solve the two-point boundary value problem:

$$\begin{bmatrix} \dot{x}^* \\ \dot{p}^* \end{bmatrix} = \begin{bmatrix} 0 & 1 & | & 0 & 0 \\ 0 & 0 & | & 0 & -1 \\ \hline -1 & 0 & | & 0 & 0 \\ 0 & -1 & | & -1 & 0 \end{bmatrix}\begin{bmatrix} x^* \\ p^* \end{bmatrix}$$

$$x^*(0) = [1 \quad 0]^T$$

$$p^*(1) = [0 \quad 0]^T \ .$$

The optimal control function is then

$$u^* = -p^{*T}\begin{bmatrix} 0 \\ 1 \end{bmatrix},$$

provided it lies in U. Solutions of two-point boundary value problems are usually not easy to obtain.

7.4 A Variational Method for Discrete-Time Systems

We next discuss the discrete-time setting. Let the control equation of the state-space description be

$$x_{k+1} = f(x_k, u_k, k) \tag{7.17}$$

with initial state $x_{k_0} = y_0$. The problem is to minimize the cost functional

$$F(\{u_k\}) = \sum_{k=k_0}^{k_1} g(x_k, u_k, k) , \tag{7.18}$$

where $\{x_k\} \in X$ and $\{u_k\} \in U$. Assuming that f and g are continuous and have continuous first partial derivatives with respect to all components of x_k and u_k and that U contains "delta sequences" of p-vectors with length $k_1 - k_0 + 1$, i.e. $\{0, \ldots, 0, y_k, 0, \ldots, 0\}$ where $y_k \neq 0$ for all $k = k_0, \ldots, k_1$, we have the following result.

Theorem 7.2 *A necessary condition for the pair* $(\{u_k^*\}, \{x_k^*\})$ *to satisfy*

$$F(\{u_k^*\}) = \min\{F(\{u_k\}): \{u_k\} \in U\} ,$$

$$x_{k+1}^* = f(x_k^*, u_k^*, k) ,$$

$$x_{k_0}^* = y_0$$

where $F(\{u_k\})$ *is given by* (7.18) *with initial state* $x_{k_0} = y_0$ *and fixed terminal time such that* $(\{u_k\}, \{x_k\})$ *satisfies* (7.17), *is that there exists a costate sequence* $\{p_k\}$ *defined by*

$$p_k = \left[\frac{\partial f}{\partial x}(x_k, u_k, k)\right]^T p_{k+1} + \left[\frac{\partial g}{\partial x}(x_k, u_k, k)\right]^T$$

$$p_{k_1 + 1} = 0$$

so that the Hamiltonian

$$H(x_k, u_k, p_{k+1}, k) = g(x_k, u_k, k) + p_{k+1}^T f(x_k, u_k, k)$$

satisfies

$$\frac{\partial H}{\partial \boldsymbol{u}}(x_k^*, \boldsymbol{u}_k^*, \boldsymbol{p}_{k+1}^*, k) = 0$$

for $k = k_0, \ldots, k_1$.

The proof of this theorem is similar to that of Theorem 7.1 (Exercise 7.10).

Exercises

7.1 Consider the following controlled damped harmonic oscillator with mass 1. Let θ be the angle of the oscillator, a the damping coefficient, and ω_0 the circular frequency. Then for small values of $|\theta|$, the motion of the oscillator can be approximated by the solution of the differential equation

$$\ddot{\theta}(t) + a\dot{\theta}(t) + \omega_0^2 \theta(t) = u(t)$$

with initial angular position and velocity $\theta(0) = \theta_0$ and $\dot{\theta}(0) = \theta_1$ respectively, where $u(t)$ represents the input control at time t. Suppose that $|u(t)| \leq 1$ and we wish to bring the oscillator to rest in a minimum amount of time. Give a mathematical description of this optimal control problem.

7.2 Prove that the three optimal control problems (i.e., the Lagrange, the Mayer, and the Bolza problems) are equivalent in the sense that each one can be reformulated as the others.

7.3 Let U be the collection of all vector-valued piecewise continuous functions $\boldsymbol{u}$ with $|\boldsymbol{u}|_2 < 1$. Prove that if $\boldsymbol{k}$ is continuous and

$$\int_{t_0}^{t_1} \boldsymbol{k}^T(t) \boldsymbol{\eta}(t) \, dt = 0$$

for all $\boldsymbol{\eta} \in U$, then $\boldsymbol{k}(t) \equiv 0$.

7.4 Verify the identity (7.7).

7.5 Prove that the necessary condition $\delta_{\boldsymbol{\eta}} F(\boldsymbol{u}) = 0$ for all $\boldsymbol{\eta} \in U$ is equivalent to (7.10).

7.6 Supply the detail of the solution of the two-point boundary value problem in determining the optimal pair (u^*, x^*) of the one-dimensional example in Sect. 7.3.

7.7 Consider the one-dimensional optimal linear servomechanism problem of finding the optimal control u^* and the corresponding optimal trajectory x^* that approximates $y(t) = 1$ such that the pair (u^*, x^*) satisfies the linear system $\dot{x} = -x + u$ with initial condition $x(0) = 0$ by minimizing the cost functional

$$F(u) = \tfrac{1}{2} \int_0^1 [(x-1)^2 + u^2] \, dt \ .$$

7.8 Prove that the optimal control u^* for the linear regulator problem (7.15) with $E=0$, $R(t)$ positive definite, and $\dot{x}=A(t)x+B(t)u$, $x(t_0)=x_0$ is a linear feedback $u^*=-K(t)x^*$ with $K(t)=R^{-1}(t)B^T(t)L(t)$ where the matrix $L(t)$ is the solution of the *matrix Riccati equation*

$$\dot{L}(t)=-L(t)A(t)-A^T(t)L(t)+L(t)B(t)R^{-1}(t)B^T(t)L(t)-Q(t)$$

$$L(t_1)=0 \ .$$

(*Hint*: Let $p=L(t)x$ in solving the two-point boundary value problem.)

7.9 Let $R(t)$ be positive definite. Prove that the optimal control function u^* for the linear servomechanism problem of minimizing

$$F(u)=\tfrac{1}{2}\int_{t_0}^{t_1}[(y-v)^T Q(t)(y-v)+u^T R(t)u]\,dt \ ,$$

where y is given, $v=C(t)x$, $\dot{x}=A(t)x+B(t)u$ and $x(t_0)=x_0$, is a linear feedback $u^*=-K(t)x+R^{-1}(t)B^T(t)z$ with $K(t)=R^{-1}(t)B^T(t)L(t)$ where the matrix $L(t)$ is the solution of the matrix Riccati equation

$$\dot{L}(t)=-L(t)A(t)-A^T(t)L(t)+L(t)B(t)R^{-1}(t)B^T(t)L(t)-C^T(t)Q(t)C(t)$$

$$L(t_1)=0$$

and the vector z is the solution of the vector differential equation

$$\dot{z}=-[A(t)-B(t)R^{-1}(t)B^T(t)L(t)]^T z-C^T(t)Q(t)y$$

$$z(t_1)=0 \ .$$

(*Hint*: Let $p=L(t)x-z$ in solving the two-point boundary value problem.)

7.10 Prove Theorem 7.2.

7.11 Let R_k be positive definite and Q_k be nonnegative definite for all $k=k_0,\ldots,k_1$. Prove that the optimal control sequence $\{u_k^*\}$ for the discrete linear regulator problem of minimizing

$$F(\{u_k\})=\frac{1}{2}\sum_{k=k_0}^{k_1}\{x_k^T Q_k x_k+u_k^T R_k u_k\}$$

where $x_{k+1}=A_k x_k+B_k u_k$ and $x_{k_0}=y_0$ is a linear feedback sequence $u_k^*=-R_k^{-1}B_k^T L_{k+1}x_k$ where the sequence $\{L_k\}$ is the solution of the matrix difference equations

$$L_k=A_k^T L_{k+1}A_{k-1}-Q_k B_{k-1}R_{k-1}^{-1}B_{k-1}^T L_k-A_k^T L_{k+1}B_{k-1}R_{k-1}^{-1}B_{k-1}^T L_k$$

$$+Q_k A_{k-1}$$

$$L_{k_1+1}=0,\quad k=k_1,\ldots,k_0+1 \ .$$

(*Hint*: Let $p_k=L_k x_{k-1}$ in solving the two-point boundary value problem.)

8. Dynamic Programming

In the previous chapter, in order to be able to apply classical variational methods, the cost functional was assumed to be differentiable with respect to each control coordinate, and to simplify the optimal control problem, we also assumed that the terminal time was fixed. In this chapter, we will drop these restrictive and very undesirable assumptions. In order to handle the more general optimal control problem, we will introduce two commonly used methods, namely: the method of *dynamic programming* initiated by Bellman, and the *minimum principle* of Pontryagin.

8.1 The Optimality Principle

As usual, we first consider the continuous-time setting. Recall that J denotes the time interval, X a subset of $\mathbb{R}^n$ to which the trajectory is confined, and U the collection of all admissible control functions. We now consider subsets of these three sets. We require the terminal time to lie in a closed sub-interval J_T of J, and the terminal state (or target) to lie in a closed subset X_T of X. Of course J_T and X_T may be singletons, and $M_T = J_T \times X_T$ will be called the *target*. For each $(\tau, y) \in J \times X$, let $U(\tau, y)$ be the subclass of control functions u in U such that the corresponding trajectory $x = x(t)$ defined by

$$\dot{x} = f(x, u, t)$$

$$x(\tau) = y$$

lies in X when $\tau \le t \le t_1$, for some terminal time $t_1 = t_1(u) \in J_T$ such that the corresponding terminal state $x(t_1)$ lies in X_T. We call $U(\tau, y)$ the admissible class of control functions with initial time-space (τ, y) (and target M_T). Note that $U(\tau, y)$ may be an empty collection. The optimal control problem is to determine an optimal control function u^* and its corresponding optimal trajectory $x^* = x^*(t)$, $t_0 \le t \le t_1^*$, where $t_1^* = t_1^*(u^*) \in J_T$ is called the corresponding (*optimal*) *terminal time*, such that $u^* \in U(t_0, x_0)$ and

$$\int_{t_0}^{t_1^*} g(x^*, u^*, t)dt = \min\left\{\int_{t_0}^{t_1} g(x, u, t)dt: u \in U(t_0, x_0)\right\}, \tag{8.1}$$

where both pairs (u^*, x^*) and (u, x) satisfy

$$\dot{x} = f(x, u, t)$$

$$x(t_0) = x_0 \tag{8.2}$$

and $t_1 = t_1(u)$ is always assumed to lie in J_T and varies with $u \in U(t_0, x_0)$. Note that if J_T is a singleton and $M_T = \mathbb{R}^n$, this problem reduces to (7.4).

The method of (continuous-time) dynamical programming depends on the following so-called "optimality principle".

Lemma 8.1 *Let (u^*, x^*) be a pair of optimal control and trajectory with initial time and state t_0 and x_0 and terminal time $t_1^* \in J_T$ for the optimal control problem (8.1–2). Then for any $\tau, t_0 < \tau < t_1^*, (u^*, x^*)$ is also an optimal control and trajectory pair with initial time-space $(\tau, x^*(\tau))$.*

To prove this lemma, we assume, on the contrary, that there is an admissible control $\tilde{u} \in U(\tau, x^*(\tau))$ whose corresponding trajectory $\tilde{x}(t)$, $\tau \le t \le \tilde{t}_1$, where $\tilde{t}_1 = \tilde{t}_1(\tilde{u}) \in J_T$, lies in X with $\tilde{x}(\tilde{t}_1) \in X_T$, such that

$$\int_\tau^{\tilde{t}_1} g(\tilde{x}, \tilde{u}, t) dt < \int_\tau^{t_1^*} g(x^*, u^*, t) dt \ .$$

Define the pair $(\hat{u}, \hat{x})$ by:

$$\hat{u} = \begin{cases} u^*(t) & \text{if} \quad t_0 < t \le \tau \\ \tilde{u}(t) & \text{if} \quad \tau < t < \tilde{t}_1 \end{cases} \quad \text{and}$$

$$\hat{x} = \begin{cases} x^*(t) & \text{if} \quad t_0 < t \le \tau \\ \tilde{x}(t) & \text{if} \quad \tau < t < \tilde{t}_1 \end{cases} \ .$$

Then we have

$$\int_{t_0}^{\tilde{t}_1} g(\hat{x}, \hat{u}, t) dt = \int_{t_0}^{\tau} g(x^*, u^*, t) dt + \int_\tau^{\tilde{t}_1} g(\tilde{x}, \tilde{u}, t) dt$$

$$< \int_{t_0}^{\tau} g(x^*, u^*, t) dt + \int_\tau^{t_1^*} g(x^*, u^*, t) dt$$

$$= \int_{t_0}^{t_1^*} g(x^*, u^*, t) dt \ .$$

Since $(\tilde{t}_1, \hat{x}(\tilde{t}_1)) = (\tilde{t}_1, \tilde{x}(\tilde{t}_1))$ is in M_T, we have a contradiction to (8.1). This completes the proof of the lemma.

8.2 Continuous-Time Dynamic Programming

An important idea of Bellman is the introduction of the extended real-valued function

$$V(\tau, y) = \min\left\{\int_\tau^{t_1} g(x, u, t)dt: u \in U(\tau, y)\right\},$$

where $t_1 = t_1(u)$, $\dot{x} = f(x, u, t)$, $x(\tau) = y$, $x(t) \in X$ for $\tau \leq t \leq t_1$, $(t_1, x(t_1))$ lies in M_T, and it is understood that $V(\tau, y) = +\infty$ if $U(\tau, y)$ is empty. $V(\tau, y)$ will be called a *value function*.

In order to establish the method of dynamic programming, we also need the following lemma which is also called an optimality principle, but will leave its proof to the reader (Exercise 8.2).

Lemma 8.2 *Let $x^*(t)$, $t_0 \leq t \leq t_1^*$, be an optimal trajectory for the optimal conrol problem (8.1, 2). Then for any t and τ with $t_0 \leq t < \tau < t_1^*$,*

$$\min_{u \in U(t, \, x^*(t))}\left\{\int_t^\tau g(x, u, s)ds + \int_\tau^{t_1} g(x, u, s)ds\right\}$$

$$= \min_{u \in U(t, \, x^*(t))}\left\{\int_t^\tau g(x, u, s)ds + \min_{\tilde{u} \in U(\tau, \, x(\tau))}\int_\tau^{\tilde{t}_1} g(\tilde{x}, \tilde{u}, s)ds\right\}.$$

It should be noted that in the last minimization process, the admissible control function $\tilde{u}$ has the initial time-space $(\tau, x(\tau))$ where x is governed by $u \in U(t, x^*(t))$. Hence, the two minimization processes on the right-hand side cannot be separated. We again remind the reader that the subscript 1 of t_1 and $\tilde{t}_1$ tells us that t_1 and $\tilde{t}_1$ are in the target: $t_1, \tilde{t}_1 \in J_T$.

The method of continuous-time dynamic programming can be summarized in the following.

Theorem 8.1 *If (u^*, x^*) exists as a pair of optimal control and trajectory with initial time-space (t_0, x_0) and terminal time $t_1^* \in J_T$ for the problem (8.1, 2), then (u^*, x^*) must satisfy both*

$$\frac{\partial V}{\partial t}(t, x^*) + \left[\frac{\partial V}{\partial x}(t, x^*)\right]f(x^*, u^*, t) + g(x^*, u^*, t) = 0, \quad t_0 \leq t \leq t_1^*$$

$$V(t_1^*, x^*(t_1^*)) = 0 \tag{8.3}$$

and

$$g(x^*, u^*, t) + \left[\frac{\partial V}{\partial x}(t, x^*)\right]f(x^*, u^*, t)$$

$$= \min_{u \in U(t, \, x^*(t))}\left\{g(x^*, u, t) + \left[\frac{\partial V}{\partial x}(t, x^*)\right]f(x^*, u, t)\right\}. \tag{8.4}$$

The first order partial differential equation (8.3) that $V(t, x)$ satisfies for $(u, x) = (u^*, x^*)$ is usually called the *Hamilton–Jacobi–Bellman equation*. To prove this theorem, let $\varepsilon > 0$ such that $t_0 \leq t < t + \varepsilon < t_1^*$. Then applying Lemma 8.1, we have

$$V(t + \varepsilon, x^*(t + \varepsilon)) - V(t, x^*(t))$$

$$= - \int_t^{t+\varepsilon} g(x^*, u^*, s)ds = - \varepsilon g(x^*, u^*, t) + o(\varepsilon) . \tag{8.5}$$

On the other hand, we also have

$$V(t + \varepsilon, x^*(t + \varepsilon)) - V(t, x^*(t)) = [V(t + \varepsilon, x^*(t + \varepsilon)) - V(t + \varepsilon, x^*(t))]$$

$$+ [V(t + \varepsilon, x^*(t)) - V(t, x^*(t))]$$

$$= \varepsilon \left[\frac{\partial V}{\partial x}(t, x^*(t)) \right] \dot{x}^*(t) + \varepsilon \frac{\partial V}{\partial t}(t, x^*(t)) + o(\varepsilon)$$

$$= \varepsilon \left\{ \left[\frac{\partial V}{\partial x}(t, x^*(t)) \right] f(x^*, u^*, t) + \frac{\partial V}{\partial t}(t, x^*(t)) + o(1) \right\} .$$

Since

$$V(t_1^*, x^*(t_1^*)) = \min \left\{ \int_{t_1^*}^{t_1} g(x, u, t)dt : u \in U(t_1^*, x^*(t_1^*)) \right\}$$

$$= \int_{t_1^*}^{t_1^*} g(x^*, u^*, t)dt = 0 ,$$

the above estimate combined with (8.5) yields the Hamilton–Jacobi–Bellman equation (8.3).

To verify (8.4), we again apply Lemma 8.1 to obtain, for $t_0 \leq t \leq t_1^*$,

$$V(t, x^*(t)) = \int_t^{t_1^*} g(x^*, u^*, s)ds$$

$$= \min_{u \in U(t, x^*(t))} \left\{ \int_t^{t+\varepsilon} g(x, u, s)ds + \int_{t+\varepsilon}^{t_1} g(x, u, s)ds \right\} .$$

Hence, using Lemma 8.2, we have

$$V(t, x^*(t)) = \min_{u \in U(t, x^*(t))} \left\{ \int_t^{t+\varepsilon} g(x, u, s)ds + V(t + \varepsilon, x(t + \varepsilon)) \right\}$$

$$= \min_{u \in U(t, x^*(t))} \{ \varepsilon g(x^*, u, t) + V(t + \varepsilon, x(t + \varepsilon)) + o(\varepsilon) \} . \tag{8.6}$$

Since

$$V(t+\varepsilon, x(t+\varepsilon)) = V(t, x(t)) + \varepsilon \left[\frac{\partial V}{\partial x}(t, x(t)) \right] f(x, u, t) + \varepsilon \frac{\partial V}{\partial t}(t, x(t)) + o(\varepsilon)$$

and $x(t) = x^*(t)$ is the initial state under the minimization process in (8.6), we may deduce from (8.6):

$$-\frac{\partial V}{\partial t}(t, x^*(t)) = \min_{u \in U(t, x^*(t))} \left\{ g(x^*, u, t) + \left[\frac{\partial V}{\partial x}(t, x^*(t)) \right] f(x^*, u, t) + o(1) \right\} .$$

Now, taking the limit as $\varepsilon \to 0$ and applying (8.3), we obtain (8.4).

Remark 8.1   To apply the method of continuous-time linear programming to determine the pair (u^*, x^*) of optimal control and trajectory, the first step is to solve for $f(x^*, u^*, t)$ and $g(x^*, u^*, t)$ in terms of $(\partial V/\partial x)(t, x^*(t))$ in the minimization process (8.4). Usually this requires writing u^* in terms of x^* and the n components of $(\partial V/\partial x)(t, x^*(t))$. If $g(x^*, u, t)$ is not differentiable with respect to the p control coordinates of u, classical variational methods cannot be applied and other "non-smooth" optimization methods are employed. The next step is to put $f(x^*, u^*, t)$ and $g(x^*, u^*, t)$, which are now in terms of (the components of) $(\partial V/\partial x)(t, x^*(t))$, or u^* in terms of x^* and $(\partial V/\partial x)(t, x^*(t))$, into (8.3) and solve this Hamilton–Jacobi–Bellman equation for $V(t, x^*)$ (usually in terms of x^*). Finally, determine (u^*, x^*) from the available information.

To illustrate this process, we return to the one-dimensional example:

$$\begin{cases} \text{minimize } \frac{1}{2} \int_0^1 [x^2(t) + u^2(t)] dt \\ \dot{x} = x + u, \qquad x(0) = 1 \end{cases}$$

discussed in Sect. 7.3. Here, since $g(x^*, u) = x^{*2} + u^2$ is smooth in u, we can simply use calculus to determine u^* in terms of x^* and $(\partial V/\partial x) = (\partial V/\partial x)(t, x^*)$ by minimizing $\frac{1}{2}(x^{*2} + u^2) + (\partial V/\partial x)(x^* + u)$, yielding

$$u^* = -\frac{\partial V}{\partial x} .$$

Thus, the Hamilton–Jacobi–Bellman equation becomes

$$\begin{cases} \frac{\partial V}{\partial t} + x^* \frac{\partial V}{\partial x} - \frac{1}{2} \left(\frac{\partial V}{\partial x} \right)^2 + \frac{1}{2} x^{*2} = 0 \\ V(1, x^*(1)) = 0 . \end{cases} \tag{8.7}$$

Observing that the term x^{*2} must be isolated, we write $V(t, x) = c(t)x^2$, so that

$$\frac{\partial V}{\partial t} = \frac{\partial V}{\partial t}(t, x^*) = \dot{c}(t)x^{*2} ,$$

and (8.7) can be simplified to give

$$\dot{c}(t) = 2c^2(t) - 2c(t) - \tfrac{1}{2}$$

$$c(1) = 0 .$$

This is the Ricatti equation (Exercises 8.5, 6). By setting $c(t) = -\dot{z}(t)/2z(t)$, we have a second order linear differential equation

$$\ddot{z} + 2\dot{z} - z = 0$$

with $\dot{z}(1) = 0$. If we pick $z(1) = 1$, we obtain

$$z(t) = \frac{1 + \sqrt{2}}{2\sqrt{2}} e^{(1 - \sqrt{2})(1 - t)} + \frac{-1 + \sqrt{2}}{2\sqrt{2}} e^{(1 + \sqrt{2})(1 - t)} ,$$

so that

$$V(t, x) = -\frac{1}{2} \frac{e^{-\sqrt{2}(1 - t)} - e^{\sqrt{2}(1 - t)}}{(\sqrt{2} + 1)e^{-\sqrt{2}(1 - t)} + (\sqrt{2} - 1)e^{\sqrt{2}(1 - t)}} x^2 .$$

Hence, we can find u^*, and then x^* by solving

$$\dot{x}^* = x^* + u^*$$

$$x^*(0) = 1 .$$

The answer for (u^*, x^*) can be shown to agree with the one obtained by using the variational approach and solving a two-point boundary value problem given in Sect. 7.3. We leave the detail to the reader (Exercise 8.4).

As mentioned in Remark 8.1, even if $g(x, u, t)$ is not smooth in u, the method of dynamic programming is still applicable. One example is the minimum-fuel control problem (Exercise 8.11).

8.3 Discrete-Time Dynamic Programming

We next consider discrete-time dynamic programming. The problem can be formulated as the following:

$$\text{minimize} \left\{ \sum_{k = k_0}^{k_1(\{u_j\})} g(x_k, u_k, k) : \{u_j\} \in U(k_0, x_0) \right\}$$

$$x_{k+1} = f(x_k, u_k, k), \quad x_{k_0} = x_0 , \tag{8.8}$$

where it is understood, as in the continuous-time setting, that (k_1, x_{k_1}), where $k_1 = k_1(\{u_j\})$, is in the time-space target M_T, and that $x_k \in X$ for $k_0 \leq k \leq k_1$. Also, the subscript 1 of k_1 always indicates that the terminal time k_1 is in the time target J_T and remember that k_1 depends on the control sequence $\{u_j\}$. Finally, $\{u_j^*\}$, $\{x_k^*\}$, and k_1^* will denote, respectively, an optimal control sequence, its corresponding optimal trajectory, and the (optimal) terminal time with initial time and state k_0 and x_0. As in the continuous-time case, we define a value function

$$V(l, y) = \min\left\{\sum_{k=l}^{k_1} g(x_k, u_k, k): \{u_k\} \in U(l, y)\right\}$$

where $x_{k+1} = f(x_k, u_k, k)$ and $x_l = y$.

The following so-called "discrete-time optimality principle" can be easily verified (Exercise 8.8).

Lemma 8.3   *For each* $l \geq k_0$,

$$V(l, x_l) = \min\left\{\sum_{k=l}^{k_1} g(x_k, u_k, k): \{u_k\} \in U(l, x_l)\right\}$$

$$= \min_{u_l}\left\{g(x_l, u_l) + \min\left[\sum_{k=l+1}^{k_1} g(\tilde{x}_k, \tilde{u}_k, k): \right.\right.$$

$$\left.\left. \{\tilde{u}_k\} \in U(l+1, f(x_l, u_l))\right]\right\}.$$

Hence, the procedure of discrete-time dynamic programming follows immediately (Exercise 8.8):

Theorem 8.2   *Let* $x_{k_0} = x_0$. *Then*

$$V(k_0, x_{k_0}) = \min_{u_{k_0}}\left\{g(x_{k_0}, u_{k_0}, k_0) + \min_{u_{k_0+1}}\left\{g(x_{k_0+1}, u_{k_0+1}, k_0+1)\right.\right.$$

$$\left.\left. + \ldots + \min_{u_{k_1}} g(x_{k_1}, u_{k_1}, k_1)\right\} \ldots\right\},$$

where $x_{k_0+1} = f(x_{k_0}, u_{k_0}), \ldots, x_{k_1} = f(x_{k_1-1}, u_{k_1-1})$.

Remark 8.2   To carry out the procedure of discrete-time programming, we pick any arbitrary k_1 and carry out the minimization processes starting with

$$\min_{u_{k_1}} g(x_{k_1}, u_{k_1}, k_1) .$$

It is important to remember that when each minimum is taken, the previous minimum quantities must be included. At the end, we have a recurrence relationship on $\{x_k\}, k = k_0, \ldots, k_1$. Suppose that $g(x_k, u_k, k)$ is nonnegative for each k. Then the smallest k_1 such that $k_1 \in J_T$ and x_{k_1} is in X_T is denoted by k_1^*, and the trajectory $x_k = x_k^*$, $k = k_0, \ldots k_1^*$, is an optimal trajectory. From $\{x_k\}$ we can determine $u_k = u_k^*$. Hence, $(\{u_k^*\}, \{x_k^*\})$, $k = k_0, \ldots, k_1^*$, is a pair of optimal control sequence and trajectory of the optimal control problem.

To illustrate the procedure, we consider the discrete linear regulator problem of minimizing

$$F(\{u_k\}) = \frac{1}{2} \sum_{k=0}^{N} (x_k^2 + u_k^2) ,$$

where $x_{k+1} = ax_k + bu_k$, a and b real, and $x_0 = y_0$. For convenience, we assume that the terminal time N is fixed. Otherwise, we follow the procedure outlined in Remark 8.2. The starting point is the trivial minimization process

$$V(N, x_N) = \min_{u_N} \tfrac{1}{2} (x_N^2 + u_N^2) .$$

It is clear that to attain the minimum, we have

$$u_N = 0,$$

$$V(N, x_N) = h_0 x_N^2 ,$$

where $h_0 = \tfrac{1}{2}$. The second minimization process is

$$\min_{u_{N-1}} \{ \tfrac{1}{2} (x_{N-1}^2 + u_{N-1}^2) + V(N, x_N) \} .$$

From Lemma 8.3, this quantity happens to be $V(N-1, x_{N-1})$. It is important to remember that $V(N, x_N)$ must be expressed in terms of u_{N-1} before the minimization is taken. That is, the second minimization process becomes:

$$V(N-1, x_{N-1}) = \min_{u_{N-1}} \{ \tfrac{1}{2} (x_{N-1}^2 + u_{N-1}^2) + h_0 (ax_{N-1} + bu_{N-1})^2 \} .$$

It is also clear that to attain the minimum, we have

$$u_{N-1} = -\frac{2abh_0}{1 + 2b^2 h_0} x_{N-1}$$

$$x_N = \frac{a}{1 + 2b^2 h_0} x_{N-1}$$

$$V(N-1, x_{N-1}) = h_1 x_{N-1}^2 , \qquad \text{where}$$

$$h_1 = \frac{1 + a^2 + b^2}{2(1 + b^2)} x_{N-1} .$$

This result suggests that $V(N-j, x_{N-j})$ is always a constant multiple of x_{N-j}^2, and so we write

$$V(N-j, x_{N-j}) = h_j x_{N-j}^2, \quad j = 1, \ldots, N .$$

Hence, the $(j+1)$st minimization process of the dynamic programming method is

$$V(N-j, x_{N-j}) = \min_{u_{N-j}} \left\{ \frac{1}{2} (x_{N-j}^2 + u_{N-j}^2) + V(N-j+1, x_{N-j+1}) \right\}$$

$$= \min_{u_{N-j}} \left\{ \frac{1}{2} (x_{N-j}^2 + u_{N-j}^2) + h_{j-1}(ax_{N-j} + bu_{N-j}) \right\}$$

and to attain the minimum, we have

$$u_{N-j} = -\frac{2abh_{j-1}}{1 + 2b^2 h_{j-1}} x_{N-j}$$

$$x_{N-j+1} = \frac{a}{1 + 2b^2 h_{j-1}} x_{N-j}$$

$$V(N-j, x_{N-j}) = h_j x_{N-j}^2 ,$$

for $j = 1, \ldots, N$. In order to determine the optimal quantity $V(0, x_0) = h_N x_0^2$, we have to find h_N. To do so, we derive its recursive relationship as in the following.

$$h_j x_{N-j}^2 = V(N-j, x_{N-j}) = \tfrac{1}{2}(x_{N-j}^2 + u_{N-j}^2) + V(N-j+1, x_{N-j+1})$$

$$= \tfrac{1}{2} (x_{N-j}^2 + u_{N-j}^2) + h_{j-1} x_{N-j+1}^2$$

$$= \frac{1}{2} \left[1 + \left(-\frac{2abh_{j-1}}{1 + 2b^2 h_{j-1}} \right)^2 + 2h_{j-1} \left(\frac{a}{1 + 2b^2 h_{j-1}} \right)^2 \right] x_{N-j}^2 ,$$

so that we have

$$h_j = \frac{1 + 2(a^2 + b^2)h_{j-1}}{2(1 + 2b^2 h_{j-1})}, \quad j = 1, \ldots, N ,$$

$$h_0 = \tfrac{1}{2} .$$

The optimal trajectory $\{x_k\} = \{x_k^*\}$ can also be computed recursively using

$$x_{N-j+1} = \frac{a}{1 + 2b^2 h_{j-1}} x_{N-j}$$

$$x_0 = y_0$$

and the optimal control sequence $\{u_k\} = \{u_k^*\}$ is now

$$u_{N-j} = -\frac{2abh_{j-1}}{1+2b^2h_{j-1}}x_{N-j} = -2bh_{j-1}x_{N-j+1} \ .$$

8.4 The Minimum Principle of Pontryagin

We have now discussed the methods of continuous-time and discrete-time
dynamic programming. Although these procedures are analogous, the
continuous-time setting involves solution of a first order nonlinear partial
differential equation. A standard method is the so-called "method of character-
istics" (see, for example, Courant and Hilbert (1962)). It is, however, usually more
preferable to solve an ordinary differential equation. This is indeed possible if we
use the *minimum principle of* Pontryagin instead. These two methods for the
continuous-time setting are very much related. In fact, under the additional
assumption that cost functionals have continuous second partial derivatives
with respect to t and the coordinates of x, we can derive Pontryagin's minimum
principle using dynamic programming.

Suppose that the Hamilton–Jacobi–Bellman equation (8.3) is satisfied. Then
denoting

$$q(t) = \left[\frac{\partial V}{\partial x}(t, x^*)\right]^T \ ,$$

we have, from (8.3),

$$\dot{q}(t) = \frac{\partial}{\partial t}\left[\frac{\partial V}{\partial x}(t, x^*)\right]^T + \frac{\partial}{\partial x}\left[\frac{\partial V}{\partial x}(t, x^*)\right]^T \dot{x}^*(t)$$

$$= \left[\frac{\partial}{\partial x}\frac{\partial V}{\partial t}(t, x^*)\right]^T + \left[\frac{\partial^2 V}{\partial x^2}(t, x^*)\right]^T f(x^*, u^*, t)$$

$$= -\left[\frac{\partial}{\partial x}\left[\frac{\partial V}{\partial x}(t, x^*)f(x^*, u^*, t) + g(x^*, u^*, t)\right]\right]^T$$

$$\quad + \left[\frac{\partial^2 V}{\partial x^2}(t, x^*)\right]^T f(x^*, u^*, t)$$

$$= -\left[\frac{\partial V}{\partial x}(t, x^*)\frac{\partial f}{\partial x}(x^*, u^*, t)\right]^T - \left[\frac{\partial g}{\partial x}(x^*, u^*, t)\right]^T$$

$$= -\left[\frac{\partial f}{\partial x}(x^*, u^*, t)\right]^T q(t) - \left[\frac{\partial g}{\partial x}(x^*, u^*, t)\right]^T \ ,$$

and $q(t_1^*)=0$ for some $t_1^* \in J_T$. Comparing with (7.11), we have $q(t)=p^*(t)$. Furthermore, if we define the Hamiltonian

$$H(x, u, p, t)=g(x, u, t)+p^T(t)f(x, u, t) , \qquad (8.9)$$

then (8.4) is equivalent to

$$H(x^*, u^*, p^*, t)= \min_{u \in U(t, x^*(t))} H(x^*, u, p^*, t), \qquad t_0 \le t \le t_1^* .$$

This is just a simplified statement of the Pontryagin's minimum principle. We summarize this in the following:

Theorem 8.3   *A necessary condition for the pair (u^*, x^*) to satisfy*

$$\begin{cases} F(u^*)=\min \{F(u): u \in U(t_0, x_0)\}, \quad F(u)=\int_{t_0}^{t_1} g(x, u, t)dt \\[2ex] \dot{x}^*=f(x^*, u^*, t), \quad t_0 \le t \le t_1 , \\[2ex] x^*(t_0)=x_0 , \end{cases}$$

where the initial condition (t_0, x_0) and the terminal condition (target) $M_T=J_T \times X_T$ are both given, is the existence of a costate p that satisfies the terminal value problem

$$\dot{p}=\left[\frac{\partial f}{\partial x}(x^*, u^*, t)\right]^T p-\left[\frac{\partial g}{\partial x}(x^*, u^*, t)\right]^T$$

$$p(t_1^*)=0 \qquad for \; some \qquad t_1^* \in J_T ,$$

such that

$$H(x^*, u^*, p^*, t)= \min_{u \in U(t, x^*(t))} H(x^*, u, p^*, t)$$

where $t_0 \le t \le t_1^*$, and the Hamiltonian $H(x, u, p, t)$ is defined in (8.9).

In the above derivation using the dynamic programming procedure we have assuemd that $g(x, u, t)$ has continuous second partial derivatives with respect to the coordinates of x. A direct proof of this theorem and a much more general result is possible under much weaker conditions on $g(x, u, t)$. We postpone discussing the more general statement of Pontryagin's principle and its discrete-time analogue to Chap. 10.

Exercises

8.1 Use the variational method discussed in Chap. 7 to solve the one-dimensional linear regulator problem

$$\text{minimize } \frac{1}{2} \int_0^2 (x^2 + u^2) dt$$

$$\dot{x} = u$$

$$x(0) = 1 \quad ,$$

and verify that $x^*(1) = (e + e^{-1})/(e^2 + e^{-2})$. Then solve the problem

$$\text{minimize } \frac{1}{2} \int_1^2 (x^2 + u^2) dt$$

$$\dot{x} = u$$

$$x(1) = (e + e^{-1})/(e^2 + e^{-2}) \quad .$$

Convince yourself of Lemma 8.1 by comparing the solutions of these two problems.

8.2 Prove Lemma 8.2.

8.3 Show that the Hamilton–Jacobi–Bellman equation for the linear regulator problem in Exercise 7.8 is

$$\frac{\partial V}{\partial t} + \frac{\partial V}{\partial x} A(t) x - \frac{1}{2} \left[\frac{\partial V}{\partial x} \right] B(t) R^{-1}(t) B^T(t) \left[\frac{\partial V}{\partial x} \right]^T + \frac{1}{2} x^T Q(t) x = 0$$

$$V(t_1, x(t_1)) = 0 \quad ,$$

and derive the matrix Riccati equation given in Exercise 7.8 by setting $V(t, x(t)) = \frac{1}{2} x^T L(t) x$.

8.4 Supply the detail of the solution in the one-dimensional example of continuous-time dynamic programming in Sect. 8.2.

8.5 Consider Riccati's equation with constant coefficients

$$\dot{x} = ax^2 + bx + c \quad , \quad a \neq 0 \quad .$$

Determine the parameter λ (in terms of a, b and c) in making the change of variable $x = \lambda \dot{z}/z$ to obtain a second order linear equation

$$\ddot{z} + \alpha \dot{z} + \beta z = 0$$

where α and β are constants in terms of a, b, and c.

8.6 Let $a(t)$, $b(t)$ and $c(t)$ be continuous functions. The first order equation

$$\dot{x} = a(t)x^2 + b(t)x + c(t)$$

is called Riccati's equation. Suppose that some particular solution x_1 of this equation is known. Show that a general solution (containing one arbitrary constant) can be obtained through the change of variable $x = (1/z) + x_1$ where z is the solution of the first order linear equation

$$\dot{z} + [b(t) + 2a(t)x_1]z + a(t) = 0 \ .$$

8.7 Apply the continuous-time dynamic programming method to solve the linear servomechanism problem

$$\text{minimize} \frac{1}{2}\int_0^1 [(x-1)^2 + u^2]dt$$

$$\dot{x} = -x + u$$

$$x(0) = 0 \ ,$$

and compare your answer with Exercise 7.7.

8.8 Prove Lemma 8.3 and use it to derive Theorem 8.2.

8.9 Use the discrete-time dynamic programming method to write a positive number r as a product of n positive numbers: $r = \prod_{i=1}^n r_i$ such that $\sum_{i=1}^n r_i$ is minimum.

(*Hint*: Let V_n be the minimum value of the sum $\sum_{i=1}^n r_i$. Then use Lemma 8.3 to establish

$$V_n = \min_{0 \le r_1 \le r} \left\{ r_1 + V_{n-1}\left(\frac{r}{r_1}\right) \right\}, \ n \ge 2) \ .$$

8.10 Apply Pontryagin's minimum principle to Exercises 7.7–9 to convince yourself that if the terminal time t_1 is fixed, $X_T = \mathbb{R}^n$, and the function $g(x, u, t)$ in the cost functional is differentiable with respect to u, then both the variational methods and Pontryagin's minimum principle give the same results.

8.11 Use Pontryagin's minimum principle to solve the one-dimensional minimum-fuel problem

$$\begin{cases} \text{minimize} \int_0^1 |u(s)|ds \ , \\ \quad {}_{u\in U} \\ U = \{u: u = \text{const}\} \ , \\ \dot{x} = x + u \ , \\ x(0) = 0, \quad x(1) = 1 \ . \end{cases}$$

9. Minimum-Time Optimal Control Problems

In Chap. 8 we derived a weaker version of Pontryagin's minimum principle using the dynamic programming procedure. A rigorous proof of the general statement of the principle is tedious. Even in the minimum-time optimal control problem where the cost functional is simply $(t_1 - t_0)$, an easy proof of the principle is not available without using functional analysis. In this chapter we will study the minimum-time optimal control problem for a continuous-time linear system in some detail and derive the minimum principle for this setting. In order to give a rigorous and yet somewhat elegant treatment, it is necessary to use some terminology and results from measure theory and functional analysis. Our original intention of presenting an elementary treatment of the subject matter is maintained if the reader is willing to accept two existence results (namely: Lemma 9.1 and the last portion of the proof of Theorem 9.2), consider "measurable functions" as "piecewise continuous functions", regard the "almost everywhere" notion as the weaker notion "everywhere with an exception of a finite number of points", and assume a set E with positive measure to be a nonempty interval.

9.1 Existence of the Optimal Control Function

The minimum-time optimal control problem for a linear system we will consider can be stated as follows:

$$\underset{u \in W}{\text{minimize}} \int_{t_0}^{t_1} 1 \, dt = \underset{u \in W}{\text{minimize}} \, (t_1 - t_0)$$

$$\dot{x} = A(t)x + B(t)u \ , \tag{9.1}$$

$$x(t_0) = x_0, \quad x(t_1) = x_1 \ ,$$

where the initial pair (t_0, x_0) and the target position x_1 are fixed, and the admissible class W consists of control functions $u = [u_1 \ \ldots \ u_p]^T$ with u_i measurable on $[t_0, \infty)$ and $|u_i| \leq 1$ almost everywhere, $i = 1, \ldots, p$. Clearly, t_1 is a function of u in the minimization process.

In order to consider a nontrivial problem, we will always assume that the state vector x can be brought from the initial position x_0 to the target position x_1 in a finite amount of time using a certain control function from W. Hence, the existence of the minimum time t_1^* [that is, $t_1^* - t_0$ is the minimum value of the extremal problem (9.1)] is trivial. The minimum-time optimal control problem we consider here is to study the existence, uniqueness, and characterization of a control function $u^* \in W$ which will be called an *optimal (minimum-time) control function*, such that

$$\dot{x} = A(t)x + B(t)u^*, \quad t_0 \leq t \leq t_1^* ,$$

$$x(t_0) = x_0, \quad x(t_1^*) = x_1 .$$

(9.2)

To facilitate the study of this problem, we introduce the notation

$$R_t = \left\{ \int_{t_0}^t \Phi(t_0, s) B(s) u(s) \, ds: \ u \in W \right\} \qquad \text{and} \qquad (9.3)$$

$$X_t = \Phi(t, t_0)[x_0 + R_t] = \left\{ \Phi(t, t_0)x_0 + \int_{t_0}^t \Phi(t, s) B(s) u(s) \, ds: \ u \in W \right\} \qquad (9.4)$$

where $\Phi(t, s)$ is the transition matrix of the linear system. We first note that these two sets have the following convenient properties.

Lemma 9.1 *For each $t \geq t_0$, R_t and X_t are both closed, bounded, and convex sets in $\mathbb{R}^n$.*

Since X_t is an affine translate of R_t in $\mathbb{R}^n$, it is sufficient to verify that R_t has the above mentioned properties. An elementary proof that R_t is closed in $\mathbb{R}^n$ is complicated. In order not to go into much detail, we apply a result from functional analysis. Let $t \geq t_0$ be fixed. To prove that R_t is closed and bounded, it is equivalent to show that it is compact. Since W is the unit ball in the product space $L_\infty[t_0, t_1] \times \cdots \times L_\infty[t_0, t_1]$ of almost everywhere bounded functions, it is "w^*-compact" and convex by the Banach-Alaoglu theorem, and hence R_t, the image of W under the transformation

$$K(u) = \int_{t_0}^t \Phi(t_0, s) B(s) u(s) \, ds, \quad u \in W , \qquad (9.5)$$

is a compact convex set in $\mathbb{R}^n$.

We are now ready to study the existence of the optimal control function u^*.

Theorem 9.1 *There exists an optimal control function $u^* \in W$ satisfying (9.2).*

From the definition of t_1^*, there exists a sequence $\{t_1^k\}$ that converges to t_1^* from above such that

$$\dot{x} = A(t)x + B(t)u_k, \quad t_0 \leq t \leq t_1^k ,$$

$$x(t_0) = x_0, \quad x(t_1^k) = x_1 , \qquad (9.6)$$

for some $u_k \in W$. The transition equation of (9.6) is

$$x_1 = \Phi(t_1^k, t_0)x_0 + \int_{t_0}^{t_1^k} \Phi(t_1^k, s)B(s)u_k(s)\,ds, \quad \text{for} \quad k = 1, 2, \ldots .$$

Let x_1^k denote the solution of (9.6); that is, $x_1^k(t)$, $t_0 \le t \le t_1^k$, is the trajectory corresponding to u_k. It is easy to see that $x_1^k(t_1^*) \to x_1$ as $k \to \infty$. Indeed, using the notation $|\cdot| = |\cdot|_2$ for the "length" of vectors (Remark 6.3), we have

$$|x_1 - x_1^k(t_1^*)| = |x_1^k(t_1^k) - x_1^k(t_1^*)|$$

$$\le |\Phi(t_1^k, t_0)x_0 - \Phi(t_1^*, t_0)x_0| + \left| \int_{t_0}^{t_1^k} \Phi(t_1^k, s)B(s)u_k(s)\,ds \right.$$

$$\left. - \int_{t_0}^{t_1^*} \Phi(t_1^*, s)B(s)u_k(s)\,ds \right|$$

$$\le |\Phi(t_1^k, t_0)x_0 - \Phi(t_1^*, t_0)x_0| + \left| \int_{t_0}^{t_1^*} [\Phi(t_1^k, s) - \Phi(t_1^*, s)]\,B(s)u_k(s)\,ds \right|$$

$$+ \left| \int_{t_1^*}^{t_1^k} \Phi(t_1^k, s)B(s)u_k(s)\,ds \right|$$

$$\le |\Phi(t_1^k, t_0) - \Phi(t_1^*, t_0)|\,|x_0| + |\Phi(t_1^k, t_1^*) - I| \int_{t_0}^{t_1^*} |\Phi(t_1^*, s)B(s)u_k(s)|\,ds$$

$$+ \int_{t_1^*}^{t_1^k} |\Phi(t_1^k, t_0)B(s)u_k(s)|\,ds$$

and this estimate tends to zero as $k \to \infty$, since $\Phi(t, t_0)$ is bounded and continuous on $[t_0, \infty)$ and each component of u_k is bounded almost everywhere by 1. It is also clear that $x_1^k(t_1^*) \in X_{t_1^*}$ where $X_{t_1^*}$ is defined by (9.4). Since $X_{t_1^*}$ is a closed set by Lemma 9.1, we may conclude that the target point x_1 is in $X_{t_1^*}$. That is,

$$x_1 = \Phi(t_1^*, t_0) + \int_{t_0}^{t_1^*} \Phi(t_1^*, s)B(s)u^*(s)\,ds$$

for some $u^* \in W$. This completes the proof of the theorem.

9.2 The Bang-Bang Principle

To study the characterization of the optimal control function u^*, let us introduce the class of so-called *bang-bang control functions* defined by

$$W_{bb} = \{u = [u_1 \ldots u_p]^T \in W : |u_i(t)| = 1 \text{ almost everywhere}, \quad i = 1, \ldots, p\}$$

and the corresponding subset

$$B_t = \{\Phi(t, t_0)x_0 + \int_{t_0}^{t} \Phi(t, s)B(s)u(s)\,ds : u \in W_{bb}\}$$

of X_t.

The following result, which is usualy called the *bang-bang principle*, essentially says that if a target position can be reached by using some admissible control function from W at $t = t_1 > t_0$, then it can also be reached by using a bang-bang control function $u \in W_{bb}$ at $t = t_1$.

Theorem 9.2   *For any $t > t_0$, $X_t = B_t$.*

Since $B_t \subseteq X_t$ and $X_t = \Phi(t, t_0)\{x_0 + R_t\}$, it is sufficient to prove that for any $y \in R_t$, where $t \geq t_0$ is fixed, there exists a bang-bang control function $\tilde{u} \in W_{bb}$ such that

$$y = \int_{t_0}^{t} \Phi(t_0, s)B(s)\tilde{u}(s)\,ds .$$

We consider the set

$$V = V_y = \{u \in W : y = \int_{t_0}^{t} \Phi(t_0, s)B(s)u(s)\,ds\}$$

and use the notion of extreme points of V. An *extreme point* $\hat{u}$ of V is a control function $\hat{u}$ in V which cannot be written as a proper convex combination of functions in V, so that $\hat{u} \neq \frac{1}{2}u_1 + \frac{1}{2}u_2$ where $u_1, u_2 \in V$. It is sufficient to show that V contains at least one extreme point and that all extreme points of V are bang-bang control functions. Suppose that $\hat{u} \in V$ is not a bang-bang control function. Then there exist a set E of positive measure in $[t_0, t]$ and an $\varepsilon > 0$ such that $|\hat{u}_i(s)| < 1 - \varepsilon$, $s \in E$, for some component $\hat{u}_i$ of $\hat{u}$. Let us consider the linear transformation K from W to $\mathbb{R}^n$ defined in (9.5) and the subcollection W_i of control functions $u = [u_1 \ldots u_p]^T$ in W where $u_j(s) = 0$ for $t_0 \leq s \leq t$ if $j \neq i$ and $u_i(s) = 0$ if $s \notin E$, $i = 1 \ldots p$. Since W_i is a "strip" in an infinite-dimensional function space, K cannot be a one-to-one transformation of W_i into its image. That is, there exists a nontrivial $\bar{u} \in W_i$ such that $K\bar{u} = 0$. Hence, both $\hat{u}_1 = \hat{u} + \varepsilon\bar{u}$ and $\hat{u}_2 = \hat{u} - \varepsilon\bar{u}$ are in V so that $\hat{u} = \frac{1}{2}(\hat{u}_1 + u_2)$ cannot be an extreme point of V. Hence, if we could prove the existence of an extreme point in V, then Theorem 9.2 is established. The proof of this fact is complicated without using results from functional analysis. We do not intend to go into detail, except by mentioning that the existence of an extreme point of V is a consequence of the Krein-Milman Theorem [see, for example, Royden (1968) p. 207] by noting that $V = K^{-1}(\{y\})$ is a nonempty, closed, bounded, convex subset of W.

As a consequence of Theorems 9.1, 2 we have the following result.

Corollary 9.1   *There exists an optimal control function u_{bb}^* in W_{bb} that satisfies (9.2).*

9.3 The Minimum Principle of Pontryagin for Minimum-Time Optimal Control Problems

Our next goal is to obtain at least a partial characterization of u_{bb}^*. Define

$$y(t) = \Phi(t_0, t)x(t) - x_0$$

and observe that $x(t) \in X_t$ if and only if $y(t) \in R_t$. Noting that $0 \in R_{t_0}$ and $R_s \subset R_t$ whenever $s < t$, we conclude that

$$R_t = \bigcup_{t_0 \leq s \leq t} R_s .$$

Since t_1^* is the smallest t_1 such that $y_1 = y(t_1) \in R_{t_1}$, y_1 must lie on the boundary $\partial R_{t_1^*}$ of $R_{t_1^*}$ whenever $x_1 = x(t_1^*) \in X_{t_1^*}$. It follows that if $x_1 \in X_{t_1^*}$ then, since $R_{t_1^*}$ is convex, y_1 must satisfy

$$z^T y_1 \geq z^T y \tag{9.7}$$

for all $y \in R_{t_1^*}$ where z is an outer normal of $R_{t_1^*}$ at y_1. The outer normal z enables us to give the following characterization of u_{bb}^*.

Theorem 9.3 *Let $u^* \in W$ be an optimal control function of the minimization problem (9.1) with minimum time t_1^* in the sense that it satisfies (9.2). Then*

$$z^T \Phi(t_0, t)B(t)u^*(t) = \max_{u \in W} z^T \Phi(t_0, t)B(t)u(t) \tag{9.8}$$

almost everywhere on $[t_0, t_1^*]$ for some nonzero constant vector $z \in \mathbb{R}^n$. Furthermore, if each component of $z^T \Phi(t_0, t)B(t)$ is almost everywhere different from zero, then the optimal control function u^* is the bang-bang control function $\text{sgn}\{B^T(t)\Phi^T(t_0, t)z\}$.

Here and throughout, we use the notation $\text{sgn}[v_1 \ldots v_p]^T = [\text{sgn } v_1 \ldots \text{sgn } v_p]^T$ where for a real number v, $\text{sgn } v$, called the signum function of v, is defined to be 1, 0, or -1 if $v > 0$, $v = 0$ or $v < 0$, respectively.

To prove this theorem, we suppose that $u^* \in W$ is an optimal cotrol function but for· any nonzero vector z in $\mathbb{R}^n$,

$$z^T \Phi(t_0, t)B(t)u^*(t) < \max_{u \in W} z^T \Phi(t_0, t)u(t)$$

on some set $E \subset [t_0, t_1^*]$ with positive measure. Let z^T be an outer normal to the boundary of $R_{t_1^*}$ at the point y_1 and $\hat{u}$ satisfy

$$z^T \Phi(t_0, t)B(t)\hat{u}(t) = \max_{u \in W} z^T \Phi(t_0, t)B(t)u(t)$$

almost everywhere on $[t_0, t_1^*]$. Then we have

$$\int_{t_0}^{t_1^*} z^T \Phi(t_0, t) B(t) u^*(t) \, dt < \int_{t_0}^{t_1^*} z^T \Phi(t_0, t) B(t) \hat{u}(t) \, dt$$

or $z^T y_1 < z^T \hat{y}$ where

$$\hat{y} = \int_{t_0}^{t_1^*} \Phi(t_0, t) B(t) \hat{u}(t) \, dt$$

is in $R_{t_1^*}$, contradicting (9.7). Finally, it is not difficult to see that if each component of $z^T \Phi(t_0, t) B(t)$ is almost everywhere different from zero, then the optimal control function u^* which satisfies (9.8) must be $\text{sgn}\{B^T(t) \Phi^T(t_0, t) z\}$ (Exercise 9.1). This completes the proof of the theorem.

Remark 9.1   If we define a vector-valued function $q(t)$ to be the unique solution of the following equation

$$\dot{q}(t) = -A^T(t) q(t), \quad t_0 \leq t \leq t_1^* ,$$

$$q(t_0) = -z$$

(9.9)

then we have $q(t) = -\Phi^T(t_0, t) z$ and so the optimal control function in Theorem 9.3 is $u^*(t) = -\text{sgn}\{B^T(t) q(t)\}$ almost everywhere on $[t_0, t_1^*]$. Furthermore, if we define the Hamiltonian to be

$$H(x, u, q, t) = 1 + q^T(t) [A(t)x + B(t)u] ,$$

(9.10)

then (9.8) can be rewritten as

$$H(x^*, u^*, q, t) = \min_{u \in W} H(x^*, u, q, t)$$

(9.11)

almost everywhere on $[t_0, t_1^*]$. Hence, Theorem 9.3 is, in fact, a minimum principle of Pontryagin.

We demonstrate Theorem 9.3 with the following example

$$\underset{u \in W}{\text{minimize}} \, t_1$$

$$\begin{bmatrix} \dot{x}_1 \\ \dot{x}_2 \end{bmatrix} = \begin{bmatrix} 0 & 1 \\ 0 & 0 \end{bmatrix} \begin{bmatrix} x_1 \\ x_2 \end{bmatrix} + \begin{bmatrix} 0 \\ 1 \end{bmatrix} u ,$$

$$\begin{bmatrix} x_1(0) \\ x_2(0) \end{bmatrix} = \begin{bmatrix} 0 \\ 0 \end{bmatrix}, \quad \begin{bmatrix} x_1(t_1) \\ x_2(t_1) \end{bmatrix} = \begin{bmatrix} 3 \\ 1 \end{bmatrix}$$

(9.12)

where the admissible class W consists of control functions u which are measurable on $[0, \infty)$ with $|u| \leq 1$ almost everywhere.

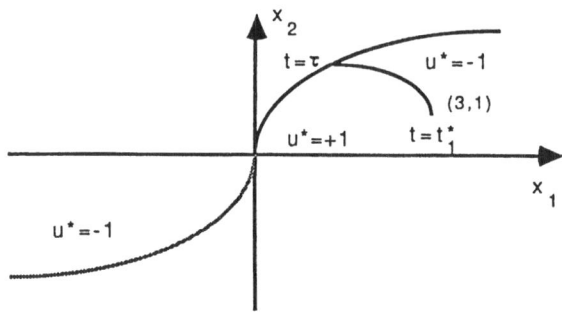

Fig. 9.1

Let u^* be an optimal control function. Then from Theorem 9.3 we have

$$u^* = -\text{sgn}\begin{bmatrix} 0 & 1 \end{bmatrix}\begin{bmatrix} q_1 \\ q_2 \end{bmatrix} = -\text{sgn}\, q_2$$

where $q_1(t) = c_1$ and $q_2(t) = -c_1 t + c_2$ for some constants c_1 and c_2 by using (9.8). We first conclude that $c_1 \neq 0$. This is clear since $c_1 = 0$ and $z \neq 0$ imply that $c_2 \neq 0$ so that u^* would be identically equal to 1 or -1, which cannot bring x from the origin to the target $(3, 1)$. Now, since $c_1 \neq 0$, q_2 has exactly one zero at $\tau = c_2/c_1$. That is, u^* changes its sign exactly once at $t = \tau$. This "break-point" is usually called the *switching time* of u^*, and it is essential since u^* cannot be identically 1 or -1.

If $u^*(t) = 1$ for $0 \leq t < \tau$, then $x_1 = \frac{1}{2}t^2$ and $x_2 = t$, which is a (half) parabola in the first quadrant of the so-called *state-phase plane*. If $u^*(t) = -1$ for $0 \leq t < \tau$, then this portion of the trajectory is in the third quadrant of this state phase plane (Fig. 9.1). Since our target position $(3, 1)$ is in the first quadrant and we are interested in minimum-time control, it is clear that we must pick $u^*(t) = 1$ for $0 \leq t < \tau$ switching to $u^*(t) = -1$ at $t = \tau$. We simply solve the two-point boundary value problem

$$\begin{bmatrix} \dot{x}_1 \\ \dot{x}_2 \end{bmatrix} = \begin{bmatrix} 0 & 1 \\ 0 & 0 \end{bmatrix}\begin{bmatrix} x_1 \\ x_2 \end{bmatrix} + \begin{bmatrix} 0 \\ 1 \end{bmatrix}(-1)$$

$$\begin{bmatrix} x_1(\tau) \\ x_2(\tau) \end{bmatrix} = \begin{bmatrix} \frac{1}{2}\tau^2 \\ \tau \end{bmatrix}, \quad \begin{bmatrix} x_1(t_1^*) \\ x_2(t_1^*) \end{bmatrix} = \begin{bmatrix} 3 \\ 1 \end{bmatrix}.$$

It is not difficult to show that the solution exists if and only if $\tau = \sqrt{14}/2$ and $t_1^* = \sqrt{14} - 1$, assuming that $0 \leq \tau \leq \tau_1^*$. Hence, the optimal control is given by

$$u^*(t) = \begin{cases} 1, & 0 \leq t < \sqrt{14}/2 \\ -1, & \sqrt{14}/2 \leq t \leq \sqrt{14} - 1, \end{cases}$$

and the minimum time is $t_1^* = \sqrt{14} - 1$.

9.4 Normal Systems

We next consider the special case where the linear system is time-invariant and described by

$$\dot{x} = Ax + Bu$$
$$x(t_0) = x_0 ,$$

(9.13)

A and B being $n \times n$ and $n \times p$ constant matrices, respectively. From Theorem 9.3, the optimal control function u^* of this problem is given by

$$u^*(t) = \text{sgn}\{z^T e^{-(t-t_0)A} B\}, \quad t_0 \le t \le t_1^* ,$$

(9.14)

for some $z \ne 0$ in $\mathbb{R}^n$. For u^* to be unique, it is essential that no component of the vector-valued signum function in (9.14) vanishes on interval. We need the following definition.

Definition 9.1   The continuous-time time-invariant linear system (9.13) is said to be *normal* if for every nonzero constant vector $z \in \mathbb{R}^n$ every component of the vector-valued function $z^T \exp[-(t-t_0)A] B$ has at most a finite number of zeros.

We remark that if the linear system is not normal, then for each nonzero z, at least one component of $z^T \exp[-(t-t_0)A]B$ is identically zero, so that the same component of u^* cannot be determined by using (9.8). In this case, we have a so-called "singular optimal control" problem.

For a normal linear system, we have the following.

Theorem 9.4   *Let $B = [b_1 \ldots b_p]$. Then the linear system (9.13) is normal if and only if each of the matrices $M_{Ab_j} = [b_j Ab_j \ldots A^{n-1}b_j]$, $j = 1, \ldots, p$, is of full rank.*

If the linear system (9.13) is not normal, then there exist $z \ne 0$ and j, $1 \le j \le p$, such that the function $f(t) = z^T \exp[-(t-t_0)A]b_j$ has infinitely many zeros on $[t_0, t_1]$ and must be identically zero, being an analytic function. Thus, we have

$$f^{(k)}(t) = (-1)^k z^T A^k e^{-(t-t_0)} b_j = 0$$

for all $t \in [t_0, t_1]$, $k = 0, 1, \ldots, n-1$. In particular, $f^{(k)}(t_0) = (-1)^k z^T A^k b_j = 0$ for $k = 0, 1, \ldots, n-1$, or, equivalently $z^T M_{Ab_j} = 0$. That is, M_{Ab_j} is row dependent and so is not of full rank.

Conversely, suppose that the matrix M_{Ab_j} is not of full rank for some j, $1 \le j \le p$. Then there exists a nonzero vector $z \in \mathbb{R}^n$ such that $z^T M_{Ab_j} = 0$, or

$$z^T b_j = z^T A b_j = \ldots = z^T A^{n-1} b_j = 0 .$$

Then by the Cayley-Hamilton Theorem, we have $z^T A^k b_j = 0$ for all $k \ge 0$. That is,

$f^{(k)}(t_0) = 0$ for $k = 0, 1, \ldots$. Since $f(t)$ is analytic for all t, it is identically zero, so that the linear system is not normal. This completes the proof of the theorem.

It is perhaps interesting to relate normality to controllability as follows.

Corollary 9.2   *Let the control matrix B in (9.13) have a single column. Then this system is normal if and only if it is completely controllable.*

For normal systems, we have the following uniqueness theorem.

Theorem 9.5   *If the continuous-time time-invariant linear system (9.13) is normal then the minimum-time optimal control function u^* is unique.*

We only prove the case when the matrix B has a single column and leave the general case to the reader (Exercise 9.7). Suppose that u_1^* and u_2^* are two optimal control functions and $x_1^*(t)$ and $x_2^*(t)$ are their corresponding (optimal) trajectories. Since the target position is the same for both control functions, we have

$$\int_{t_0}^{t_1^*} e^{-(t-t_0)A} B[u_1^*(t) - u_2^*(t)]\,dt = 0 \;. \tag{9.15}$$

Let z_1 be a nonzero constant vector in $\mathbb{R}^n$ so chosen that

$$u_1^*(T) = \mathrm{sgn}\{z_1^T e^{-(t-t_0)A} B\}^T$$

almost everywhere on $[t_0, t_1^*]$. Then, since $|u_2^*| \leq 1$, we must have

$$z_1^T e^{-(t-t_0)A} B[u_1^*(t) - u_2^*(t)] \geq 0 \tag{9.16}$$

so that multiplying z_1^T to the left of (9.15) gives

$$z_1^T e^{-(t-t_0)A} B[u_1^*(t) - u_2^*(t)] = 0$$

almost everywhere on $[t_0, t_1^*]$. Since the linear system is normal, the scalar-valued function $z_1^T \exp[-(t-t_0)A]B$ has at most a finite number of zeros so that $u_1^*(t) - u_2^*(t) = 0$ almost everywhere on $[t_0, t_1^*]$, establishing the uniqueness result.

When B has a single column, we have the following result that governs the numbers of switching times.

Theorem 9.6   *If the linear system (9.13) is a single-input normal continuous-time time-invariant system, then its minimum-time optimal control function u^* has a finite number of switching times. Furthermore, if all the eigenvalues of the system matrix A are real, then the number of switching times of u^* is at most $n-1$.*

Let $z \neq 0$ and $u^* = \mathrm{sgn}\{z^T \exp[-(t-t_0)A]B\}$. Since the analytic function $z^T \exp[-(t-t_0)A]B$ has only finitely many zeros on $[t_0, t_1^*]$, u^* has a finite number of switching times.

Suppose that all the eigenvalues $\lambda_1, \ldots, \lambda_n$ of A are real. Let us first assume that they are distinct. Then we may write $A = P \operatorname{diag}[\lambda_1, \ldots, \lambda_n]P^{-1}$ for some nonsingular matrix P. It follows easily (Exercise 9.8) that

$$u^* = \text{sgn}\{z^T e^{-(t-t_0)A}B\}$$

$$= \text{sgn}\{z^T P \, \text{diag}[e^{-\lambda_1(t-t_0)}, \ldots, e^{-\lambda_n(t-t_0)}]P^{-1}B\}$$

$$= \text{sgn}\left\{\sum_{j=1}^{n} c_j e^{-\lambda_j(t-t_0)}\right\},$$

where $c_1, \ldots, c_n$ are real constants. Since the polynomial

$$p(x) = \sum_{j=1}^{n} c_j x^{\lambda_j}$$

has at most $(n-1)$ positive roots by the Descarte's rule of signs, and $x = \exp[-(t-t_0)]$ is a monotone decreasing function at t, the number of zeros of $p\exp[-(t-t_0)]$ does not exceed $n-1$, so that $u^*(t) = \text{sgn}\{p\exp[-(t-t_0)]\}$ has at most $(n-1)$ switching times.

In general, suppose that the eigenvalues of A are $\mu_1, \ldots, \mu_k$ with multiplicities $m_1, \ldots, m_k$, respectively, where $m_1 + \ldots + m_k = n$. Then using (6.5) we have

$$u^*(t) = \text{sgn}\left\{z^T \sum_{j=1}^{k} \sum_{l=0}^{m_j-1} \frac{(t-t_0)^l}{l!} e^{\mu_j(t-t_0)}P_{l_j}B\right\}$$

$$= \text{sgn}\left\{\sum_{j=1}^{k} c_j(t)e^{\mu_j(t-t_0)}\right\},$$

where P_{l_j} are constant matrices and each $c_j(t)$, $j = 1, \ldots, k$, is a polynomial of degree $m_j - 1$. A mathematical induction proof (Exercise 9.9) shows that the function

$$h_k(t) = \sum_{j=1}^{k} c_j(t)e^{\mu_j(t-t_0)}$$

has at most $m_1 + \ldots + m_k - 1 = n - 1$ real zeros. This completes the proof of the theorem.

Exercises

9.1 Prove that if a vector-valued measurable function u^* satisfies

$$y^T(t)u^*(t) = \max_{u \in W} y^T(t)u(t)$$

almost everywhere on $[t_0, t_1^*]$ for some vector-valued measurable function y, where each component of y is almost everywhere different from zero and the admissible class W consists of vector-valued functions $u = [u_1 \ldots u_p]^T$ with each u_i measurable and $|u_i| \le 1$ almost everywhere, then $u^*(t) = \text{sgn}\{y(t)\}$ almost everywhere.

9.2 Let W be the class of all measurable functions u with $|u| \le 1$. Solve the minimum-time optimal control problem:

minimize t_1

$$\begin{bmatrix} \dot{x}_1 \\ \dot{x}_2 \end{bmatrix} = \begin{bmatrix} 0 & 1 \\ 0 & 0 \end{bmatrix} \begin{bmatrix} x_1 \\ x_2 \end{bmatrix} + \begin{bmatrix} 0 \\ 1 \end{bmatrix} u$$

$$\begin{bmatrix} x_1(0) \\ x_2(0) \end{bmatrix} = \begin{bmatrix} 3 \\ -1 \end{bmatrix}, \quad \begin{bmatrix} x_1(t_1) \\ x_2(t_1) \end{bmatrix} = \begin{bmatrix} 0 \\ 0 \end{bmatrix}.$$

9.3 Prove that the minimum-time optimal control function u^* for the damped harmonic oscillator discussed in Exercise 7.1 with $a^2 = 4\omega_0^2$ is given by

$$u^*(t) = \text{sgn}\{e^{at/2}(z_1 t + z_2)\}$$

where $z = [z_1 \ z_2]^T$ is an outer normal vector discussed in Theorem 9.3.

9.4 When the system is nonlinear, the corresponding minimum-time optimal control problem may not have a bang-bang solution. This can be seen in the following example. Consider the nonlinear system

$$\dot{x} = u - u^2 .$$

Show that the minimum-time optimal control using measurable functions u with $|u| \le 1$ taking x from $x_0 = 0$ to $x_1 = 1$ is the unique solution $u^* \equiv \frac{1}{2}$.

9.5 Verify that the two-dimensional system described by (9.12) is normal and the eigenvalues of the system matrix are all real and distinct so that by Theorem 9.6 the (unique) optimal control function has at most one switching time.

9.6 Determine the normality for the linear system with the system matrix A and control matrix B given by

$$A = \begin{bmatrix} 0 & 1 & 0 \\ 0 & 0 & 1 \\ 0 & 0 & 0 \end{bmatrix} \quad \text{and} \quad B = \begin{bmatrix} 0 \\ 0 \\ 1 \end{bmatrix}.$$

Also, verify that the number of switching times on $[0, \infty)$ for the corresponding optimal control function u^* is at most 2 by expressing u^* to be the signum function (9.14).

9.7 Prove Theorem 9.5 when B is an $n \times p$ arbitrary constant matrix.

9.8 Show that if $A = P \operatorname{diag}[\lambda_1, \ldots, \lambda_n] P^{-1}$ where P is a nonsingular constant matrix, then

$$e^{-(t-t_0)A} = P \operatorname{diag}[e^{-\lambda_1(t-t_0)}, \ldots, e^{-\lambda_n(t-t_0)}] P^{-1} \ .$$

9.9 Use mathematical induction to prove that the function

$$h_k(t) = \sum_{j=1}^{k} c_j(t) c^{\mu_j(t-t_0)} \ ,$$

where $\mu_1, \ldots, \mu_k$ are distinct real numbers, each $c_j(t)$ is a polynomial of degree $m_j - 1$, and $j = 1, \ldots, k$, has at most $m_1 + \ldots + m_k - 1$, positive zeros.

10. Notes and References

In our attempt to introduce the state-space approach to control theory, we have only included what we believe to be the most basic topics that give the reader a good preparation for further investigation into other areas of the subject. Our treatment has been elementary and yet mathematically rigorous. There are many texts in the literature that are written for similar but different purposes. For linear system theory, we refer the reader to Balakrishnan (1983), Brockett (1970), Chen (1984), Kailath (1980), Padulo and Arbib (1974), Timothy and Bona (1968), and Zadeh and Desoer (1979). For further investigation into optimal control theory, the reader is referred to Bellman (1962), Fleming and Rishel (1975), Knowles (1981), Lee and Markus (1967), Macki and Strauss (1982), and Pontryagin et al. (1962). It is an impossible task to list all other topics that we have not covered in this treatise. We only include the following related ones without going into details, and refer the interested reader to the appropriate literature.

10.1 Reachability and Constructibility

Recall that a linear system is said to be controllable if starting from any position x_0 in $\mathbb{R}^n$ the state vector can be brought to the origin by a certain control function in a finite amount of time (Definition 3.1). If the reverse process can be performed, the linear system is said to be *reachable*. In other words, the system is said to be reachable, if for any given target y_0 in $\mathbb{R}^n$, a control function can be chosen to bring the state vector from the origin to y_0 within a finite amount of time. Just as observability is "dual" to controllability, the "duality" of reachability is *constructibility*. More precisely, a continuous-time linear system is said to be (completely) constructible over the time interval $[t_0, t_1]$, if for any given input function $u(t)$, $t_0 \leq t \leq t_1$, the terminal state $x(t_1)$ is uniquely determined by the input-output pair $(u(t), v(t))$, $t_0 \leq t \leq t_1$. Of course, an analogous definition can easily be formulated for discrete-time linear systems. See Kailath (1980) and the references therein for more detail.

10.2 Differential Controllability

A linear system with continuous-time state-space description

$$\dot{x} = A(t)\,x + B(t)u$$

$$v = C(t)x + D(t)u$$

is said to be *differentially (completely) controllable* at time t_0, if starting from any position x_0 in $\mathbb{R}^n$, the state vector x at t_0 can be brought to any other position x_1 in $\mathbb{R}^n$ in an arbitrarily small amount of time by certain control function u. Assume that $A(t)$ and $B(t)$ are respectively $n \times n$ and $n \times p$ matrices with infinitely differentiable entries, and set

$$M_0(t) = B(t), \quad M_{k+1}(t) = -A(t)M_k(t) + \frac{d}{dt}M_k(t), \quad k = 0, 1, \dots ,$$

and

$$M_{AB}(t) = [M_0(t) \ M_1(t) \ \dots \ M_{n-1}(t) \ \dots] \ .$$

Then this system is differentially completely controllable at t_0 if and only if the matrix $M_{AB}(t_0)$ has rank n (for more detail, see Chen (1984)).

10.3 State Reconstruction and Observers

If a continuous-time linear system described by

$$\dot{x} = A(t)x + B(t)u$$

$$v = C(t)x$$

is observable, we have seen that the initial state $x(t_0)$ and hence the state vector $x(t)$, $t > t_0$, can be (uniquely) constructed, at least theoretically, from the information on the input-output pair $(u(\tau),\ v(\tau))$ for $t_0 \le \tau \le t$. In fact, from Chap. 4, we have:

$$x(t) = \Phi(t, t_0)P_t^{-1}\left[\int_{t_0}^{t} \Phi^T(\tau, t_0)C^T(\tau)v(\tau)\,d\tau \right.$$

$$\left. - \int_{t_0}^{t}\int_{t_0}^{\tau} \Phi^T(\tau, t_0)C^T(\tau)C(\tau)\Phi(\tau, s)B(s)u(s)\,ds\,d\tau \right] ,$$

where P_t is given in (4.2). However, if the system is not observable, so that P_t is singular, we need an *observer* to give an estimate $\hat{x}$ of x. One usually requires that

$|\hat{x}(t) - x(t)| \to 0$ as $t \to +\infty$. An observer is an associated system defined by

$$\dot{\hat{x}} = A(t)\hat{x} + B(t)u + G(t)[v - C(t)\hat{x}]$$

$$\hat{x}(t_0) = \hat{x}_0$$

and the problem is to "design" the *gain matrix* $G(t)$ so that the estimation satisfies the specification. Let $y = x - \hat{x}$ be the error. Then combining the observer and the original linear system description, we have

$$\dot{y} = \dot{x} - \dot{\hat{x}} = A(t)y - G(t)[v - C(t)\hat{x}]$$

$$= A(t)y - G(t)[C(t)x - C(t)\hat{x}]$$

$$= [A(t) - G(t)C(t)]y \ .$$

This is a new free linear system. Let $\Psi_G(t, s)$ be its transition matrix. By Theorem 6.3, we can conclude that the estimation satisfies the specification (i.e. $|\hat{x}(t) - x(t)| \to 0$ as $t \to +\infty$) if and only if

$$\int_s^t |\Psi_G(\tau, s)| d\tau \le M < \infty$$

for all $t \ge s \ge t_0$, provided that the matrix $A(t) - G(t)C(t)$ is bounded for all $t \ge t_0$. This is a specification on the design of the gain matrix $G(t)$. For time-invariant systems, another specification is to choose G such that all the eigenvalues of $A - GC$ lie in the left (open) half complex plane (Theorem 6.2). If the original system is already observable, the estimation could improve its exponent on exponential stability. Indeed, it is proved in Wonham (1967) and O'Reilly (1983) that *a gain matrix G exists such that the matrix $A - GC$ has arbitrarily assigned eigenvalues if and only if the observability matrix N_{CA} is of full rank.*

In some applications it is conceivable that the dimension n of the state vector x is very large. Hence, it is important to construct an estimator $\hat{x}$ with fewer state variables. The associated system that defines the estimator with the minimum number of equations is called a *minimal-order observer.* It is known that the dimension of the minimal-order observer is at most $n - q$ (cf. Luenberger (1964) and O'Reilly (1983)).

10.4 The Kalman Canonical Decomposition

The decomposition described in Theorem 5.1 was first considered in Gilbert (1963) where the eigenvalues of the system matrix were assumed to be distinct. A generalization to time-varying systems was studied in Kalman (1962, 1963) and Weiss (1969). However, we would like to point out again that as the example described by (5.3) indicates, there is no guarantee that the subsystems $\mathscr{S}_1$ and $\mathscr{S}_4$

are, respectively, completely controllable and observable, although we have arrived at the desired decomposed form. In fact, a unitary transformation cannot change the situation and a more general nonsingular transformation may be required.

The essential idea initiated in Kalman (1962, 1963) is to utilize the fact that the intersection of the null space $N_0 = vN_{cA}$ of N_{cA} and sp M_{AB} is invariant under A. To carry out this idea in more detail, the decomposition transformation matrix was formed in Sun (1984) by using certain basis of $V_1 \oplus \ldots \oplus V_4 = \mathbb{R}^n$ as columns, with $V_1 = \text{sp } M_{AB} \cap N_0$, $V_2 = \text{sp } M_{AB} \cap R_0$, $V_3 = N_c \cap N_0$, and $V_4 = N_c \cap R_0$, where $N_c \oplus \text{sp } M_{AB} = R_0 \oplus N_0 = \mathbb{R}^n$. We note, however, that the invariance of V_1 under A *alone* does not guarantee the complete controllability of the subsystem $\mathscr{S}_1$. This can be seen in the following example. Let

$$A = \begin{bmatrix} 1 & 1 & 0 \\ 0 & 1 & 1 \\ 0 & 0 & 1 \end{bmatrix}, \quad B = \begin{bmatrix} 0 \\ 0 \\ 1 \end{bmatrix}, \quad C = [0 \ 1 \ 1],$$

so that

$$M_{AB} = \begin{bmatrix} 0 & 0 & 1 \\ 0 & 1 & 2 \\ 1 & 1 & 1 \end{bmatrix}, \quad N_{cA} = \begin{bmatrix} 0 & 1 & 1 \\ 0 & 1 & 2 \\ 0 & 1 & 3 \end{bmatrix},$$

$$N_0 = \text{sp} \left\{ \begin{bmatrix} 1 \\ 0 \\ 0 \end{bmatrix} \right\}, \quad N_c = \{0\} \quad \text{and}$$

$$V_1 = \text{sp} \left\{ \begin{bmatrix} 1 \\ 0 \\ 0 \end{bmatrix} \right\}, \quad V_3 = V_4 = \{0\}.$$

By choosing $V_2 = \text{sp}\{[0 \ 0 \ 1]^T, [0 \ 1 \ 1]^T\}$, we obtain the transformation matrix

$$G = \begin{bmatrix} 1 & 0 & 0 \\ 0 & 0 & 1 \\ 0 & 1 & 1 \end{bmatrix}$$

so that

$$\tilde{A} = G^{-1}AG = \begin{bmatrix} 1 & 0 & 1 \\ 0 & 0 & -1 \\ 0 & 1 & 2 \end{bmatrix}, \quad \tilde{B} = G^{-1}B = \begin{bmatrix} 0 \\ 1 \\ 0 \end{bmatrix}, \quad \tilde{C} = CG = [0 \ 1 \ 2].$$

It is easy to see that the subsystem $\mathscr{S}_1$ is neither controllable nor observable although V_1 is the intersection of the controllable subspace sp M_{AB} and N_0 (for more detail, see Chen and Chui (1986)).

10.5 Minimal Realization

If the system, control, and observation matrices of a state-space description of a time-invariant linear system are given, the transfer function of the system can easily be calculated by using (5.11). The inverse of this problem is much more important, and many methods are available to estimate the impulse responses (6.23) or (6.29), and hence the transfer functions by using Laplace transform or z-transform, respectively. This problem which is known as the *realization* problem obviously does not have unique solutions. One would usually prefer, however, to determine a state-space description with the lowest dimensions. The solution of this so-called *minimal realization problem* is indeed "unique" (up to a similar transformation) according to Kalman (1963), if it exists; and the existence is guaranteed provided that the time-invariant linear system is both completely controllable and observable (Silverman (1971)). This important problem will be further investigated in a forthcoming monograph by the present authors.

10.6 Stability of Nonlinear Systems

We have already considered stability of a free linear system described by $\dot{x} = A(t)x$ where $A(t)$ is an $n \times n$ matrix with continuous entries. More generally, a free system may have a possibly nonlinear description:

$$\dot{x} = f(x, t) \tag{10.1}$$

where f is a vector-valued function defined on $Q \times J$, with $Q \subset \mathbb{R}^n$ and $J = [t_0, \infty)$. In applications, f must be assumed to be smooth enough that (10.1) with any initial condition has a unique solution. A point x_e in Q is called an *equilibrium point* (or *state*) if equation (10.1) with initial state $x(t_0) = x_e$ has the unique solution $x(t) = x_e$ for all $t \geq t_0$. Hence, any equilibrium point must satisfy the equation $f(x_e, t) = 0$ for all $t \geq t_0$. By the change of variable $g(x, t) = f(x + x_e, t)$, it is sufficient to consider the equilibrium point to be $x_e = 0$, and of course, we must assume that 0 is in the interior of Q. It is clear that the stability definitions in Chap. 6 are valid for this more general and possibly nonlinear situation. In the study of stability of nonlinear systems, the main tool is the so-called *Lyapunov function*.

Let $V(x, t)$ be a scalar-valued continuous function in $Q \times J$ such that each of the first partial derivatives

$$\frac{\partial V}{\partial x_1}, \cdots, \frac{\partial V}{\partial x_n}, \frac{\partial V}{\partial t}$$

is also continuous in $Q \times J$. We say that $V(x, t)$ is a *Lyapunov function*, if it satisfies the following conditions throughout $Q \times J$:

i) $V(0, t) = 0$ for all $t \geq t_0$.
ii) $V(x, t) > 0$ for all $x \neq 0$ and $t \geq t_0$,   and
iii) $(dV/dt) < 0$ for all $x \neq 0$ and $t \geq t_0$.

Here, the (total) derivative of $V(x, t)$ is given by

$$\frac{dV}{dt} = \left(\frac{\partial V}{\partial x}\right)^T \dot{x} + \frac{\partial V}{\partial t} = \left(\frac{\partial V}{\partial x}\right)^T f(x, t) + \frac{\partial V}{\partial t} \; . \tag{10.2}$$

The famous Lyapunov Theorem says that *if a Lyapunov function $V(x, t)$ exists, then the free system described by (10.1) is asymptotically stable about 0; that is, there exists a $\delta > 0$ such that whenever $|x(t_0)| < \delta$, $|x(t)| \to 0$ as $t \to +\infty$.*

This local stability result can be made global if $V(x, t)$ satisfies the additional condition

iv) $V(x, t) \to \infty$ as $|x| \to \infty$.

(This "limit" means that for any positive number M_1, there exists another positive number M_2, such that whenever $|x(t)| \geq M_2$ we have $V(x, t) \geq M_1$ for the same values of t.) The stronger statement of Lyapunov's theorem is that *if a Lyapunov function $V(x, t)$ exists and satisfies (iv), then any state x described by (10.1) must tend to 0 as $t \to +\infty$ (independent of the initial state).*

The relation of the Lyapunov function and the differential equation (10.1) is given by (iii) using (10.2).

There is also a Lyapunov instability theorem which states that *if there exists a scalar-valued continuous function $U(x, t)$ on $Q \times J$ such that all its first partial derivatives are also continuous on $Q \times J$, and that $U(x, t)$ satisfies*

i) *$U(0, t) = 0$ for all $t \geq t_0$,*
ii) *there exists a sequence $x_k \neq 0$ in Q that tends to 0 such that $U(x_k, t) > 0$ for all $t \in J$ and all k, and*
iii) *$(dU(x, t)/dt) = (\partial U/\partial x)^T f(x, t) + (\partial U/\partial t) > 0$ for $t \geq t_0$ all x in Q that are sufficiently close to but different from 0,*

then the system described by (10.1) is unstable about 0.

For non-free systems, that is, those described by

$$\dot{x} = f(x, u, t) \tag{10.3}$$

where u is the control function, an analogous (but slightly more complicated) stability result of Lyapunov can be formulated. For more details in this direction, we refer the reader to Lefschetz (1965a, 1965b).

10.7 Stabilization

Let us return to linear systems. Suppose that the free linear system $\dot{x} = A(t)x$ is unstable and we have a state-space description with the control equation $\dot{x} = A(t)x + B(t)u$. One method to stabilize the free system is to introduce a certain *linear feedback*:

$$u = K(t)x \ ,$$

such that the "free" linear system

$$\dot{x} = [A(t) + B(t)K(t)]x$$

is stable. For time-invariant systems, the following result is useful in stabilization (Willems and Miller (1971), and Wonham (1967, 1974)):

There exists a feedback matrix K, such that the eigenvalues of the matrix $A - BK$ can be arbitrarily assigned, if and only if the controllability matrix M_{AB} is of full rank.

10.8 Matrix Riccati Equations

In solving the linear regulator and servomechanism problems (Exercises 7.8, 9), we have to solve the matrix Riccati equation

$$\dot{L}(t) = -L(t)A(t) - A^T(t)L(t) + L(t)B(t)R^{-1}(t)B^T(t)L(t) - Q(t), \ t_0 \leq t \leq t_1 \ ,$$

$$L(t_1) = S$$

in order to obtain a linear feedback control function. Here, t_1 is fixed and S a constant matrix which may be zero. To solve this terminal value problem of a nonlinear matrix differential equation, we could instead solve the initial value problem

$$\begin{bmatrix} \dot{M} \\ \dot{N} \end{bmatrix} = \begin{bmatrix} A(t) & -B(t)R^{-1}(t)B^T(t) \\ -Q(t) & -A^T(t) \end{bmatrix} \begin{bmatrix} M \\ N \end{bmatrix}$$

$$\begin{bmatrix} M(t_1) \\ N(t_1) \end{bmatrix} = \begin{bmatrix} I \\ S \end{bmatrix}$$

and obtain $L(t)$ using $L=NM^{-1}$. Indeed, it is routine to check that if

$$\begin{bmatrix} M \\ N \end{bmatrix}$$

satisfies the initial value problem and M is invertible, then $L=NM^{-1}$ solves the above matrix Riccati equation. That M is actually invertible follows by observing that $M(t)=\Phi(t, t_1)$ where $\Phi(t, \tau)$ is the transition matrix of the linear system

$$\dot{M}=[A(t)-B(t)R^{-1}(t)B^T(t)L(t)] M .$$

Note also that $N(t)=L(t)\Phi(t, t_1)$ so that $L=NM^{-1}$. For more detail on this subject we refer the interested reader to Brockett (1970).

10.9 Pontryagin's Maximum Principle

The minimum principle of Pontryagin that we discussed in Chap. 8 was called the maximum principle in the original book of Pontryagin et al. (1962). Of course, a simple sign change in the costate vector p changes minimum back to maximum, namely:

$$\min H(x, u, p, t)= -\max H(x, u, -p, t)$$

(cf. (8.9)). In a more general setting, consider an optimal control problem in which the continuous-time system is described by

$$\dot{x}=f(x, u, t), \quad t\in J,$$
$$x(t_0)=x_0$$

where $x\in R^n$, $u\in \mathbb{R}^p$ with $p\le n$, and f is a continuously differentiable vector-valued function. The initial time and position $t_0 \in J$ and x_0 respectively are both given, and the problem is to bring the state vector x from x_0 to the target position $x_1 \in X_T$ with terminal time $t_1 \in J_T$, by using some admissible control function u, so that the cost functional

$$F(u)=\int_{t_0}^{t_1} g(x, u, t)\, dt$$

is minimized. Here, X_T and J_T are prescribed closed subsets of $\mathbb{R}^n$ and J, respectively, and the admissible class of control functions is

$$W=\{u\in \mathbb{R}^p: u_i \text{ measurable and } |u_i|\le 1 \text{ almost everywhere, } i=1, \ldots, p\} .$$

For technical reasons, the function $g(x, u, t)$ is assumed to be continuously differentiable with respect to each component of x. Let us define the Hamiltonian

$$H(x, u, p, p_0, t)=p_0 g(x, u, t)+p^T f(x, u, t)$$

and set

$$M(x, p, p_0, t) = \max_{u \in W} H(x, u, p, p_0, t) .$$

Then, Pontryagin's maximum principle can be stated as follows (Lee and Markus (1967), Knowles (1981), and Pontryagin et al (1962)): *If* u^* *is an optimal control function with corresponding trajectory* x^* *and terminal time* t_1^*, *then there exist nonpositive constant* p_0 *and a vector-valued continuous function* $p(t)$ $= [p_1(t) \ldots p_n(t)]^T$ *such that*

i) $\begin{cases} \dot{x}^* = \left[\dfrac{\partial H}{\partial p}(x^*, u^*, p, p_0, t) \right]^T = f(x^*, u^*, t) , \\[3mm] \dot{p} = -\left[\dfrac{\partial H}{\partial x}(x^*, u^*, p, p_0, t) \right]^T = p_0 \dfrac{\partial g}{\partial x}(x^*, u^*, t) + \left[\dfrac{\partial f}{\partial x}(x^*, u^*, t) \right] p , \end{cases}$

where $t_0 \le t \le t_1^*$,

ii) $H(x^*, u^*, p, p_0, t) = M(x^*, p, p_0, t), \; t_0 \le t \le t_1^*$,   *and*

iii) $M(x^*, p, p_0, t) = \displaystyle\int_{t_1^*}^{t} \left\{ p^T(s) \dfrac{\partial f}{\partial t}(x^*(s), u^*(s), s) + p_0 \dfrac{\partial g}{\partial t}(x^*(s), u^*(s), s) \right\} ds .$

Note that $M(x^*, p, p_0, t_1^*) = 0$.

In the discrete-time setting, let us discuss an analogous control problem where the system equation is

$$x_{k+1} = f_k(x_k, u_k), \quad k = 0, 1, \ldots, N-1 .$$

Here, for each $k = 0, \ldots, N-1$, $x_k \in \mathbb{R}^n$, $u_k \in \mathbb{R}^p$ with $p \le n$ and f_k is a continuously differentiable vector-valued function. Suppose that each $X_k \subseteq \mathbb{R}^n$, $k = 0$, $1, \ldots, N$, and $U_k \subseteq \mathbb{R}^p$, $k = 0, 1, \ldots, N-1$. Then the optimal control problem is to find a sequence $\{u_k\}$ of admissible control functions and a corresponding sequence $\{x_k\}$ of trajectories such that a given functional $F(x_N)$, such as the Pontryagin function (Sect. 7.1), say, is to be maximized, subject to the constraints $u_k \in U_k$, $k = 0, 1, \ldots, N-1$, and $x_k \in X_k$, $k = 0, 1, \ldots, N$.

A set A in $\mathbb{R}^n$ is called an *affine set* if $[(1 - \lambda)x + \lambda y] \in A$ for every $x, y \in A$ and $\lambda \in \mathbb{R}^1$, and the smallest affine set containing a set H is called the *affine hull* of H, denoted by aff H. The *relative interior* of a convex set C in $\mathbb{R}^n$ is defined to be

$$\text{ri } C = \{x \in \text{aff } C : (x + \varepsilon S) \cap (\text{aff } C) \subset C \text{ for some } \varepsilon > 0\}$$

where S is the unit ball $|x|_2 \le 1$ in $\mathbb{R}^n$. Let $x \in X \subseteq \mathbb{R}^n$. A closed convex cone C is called a *derived cone* of X at x if for any collection of vectors $p_1, \ldots, p_k$ in ri C, there exists a neighborhood B of the origin relative to $\mathbb{R}^k_+$ and a C^1 map $m: B \rightarrow X$, satisfying

$$m(\tau) = x + \sum_{i=1}^{k} \tau_i p_i + o(\tau), \quad \text{as } \tau \rightarrow 0 ,$$

where $\tau = [\tau_1, \ldots, \tau_k]^T \in B$. The discrete-time Pontryagin maximum principle can be stated as follows (Wonham (1968)): *In the above problem, let $V_k(x)$ $= f_k(x, U_k)$ be convex for every $x \in \mathbb{R}^n, \kappa = 0, 1, \ldots, N - 1$. Let the pair $\{u^*_k\}, \{x^*_k\}$ be an optimal solution of the control problem and C_k a derived cone of X_k at x^*_k, $k = 0, 1, \ldots, N$. Then there exist a number $\mu \geq 0$, and vectors $p_k, q_k, k = 0, 1, \ldots, N$, such that*

i) $p_k = \left[\dfrac{\partial f}{\partial x} (x^*_k, u^*_k) \right]^T p_{k+1} - q_k, k = 0, 1, \ldots, N-1,$

ii) $q^T_k x \leq 0$, *for all* $x \in X_k$,

iii) $p^T_{k+1} f_k(x^*_k, u^*_k)] = \max\limits_{u \in U_k} p^T_{k+1} f_k(x^*_k, u_k)$,

iv) $p_0 = 0, p_N = \mu \left[\dfrac{\partial F}{\partial x} (x_N) \right]^T - q_N$, *and*

$(\mu, p_0, \ldots, p_N, q_0, \ldots, q_N) \neq 0$.

10.10 Optimal Control of Distributed Parameter Systems

In practice, a great variety of control systems can be described by a partial differential equation

$$f\left(z, \frac{\partial z}{\partial x}, \frac{\partial z}{\partial t}, \frac{\partial^2 z}{\partial x \partial x}, \frac{\partial^2 z}{\partial x \partial t}, \frac{\partial^2 z}{\partial t^2}, u, v, w, x, t \right) = 0 ,$$

where t is the time variable restricted to $[t_0, t_1] \subset J, x = [x_1 \ldots x_m]^T$ a point in a region X, $z = [z_1(x, t) \ldots z_n(x, t)]^T$ restricted to a region Z with each $z_i(x, t)$ being a continuously differentiable function with respect to both x and t, $u = [u_1(t) \ldots u_r(t)]^T$, $v = [v_1(x) \ldots v_s(x)]^T$, $w = [w_1(x, t) \ldots w_h(x, t)]^T$ $(r + s + h \leq n)$ are vector-valued control functions belonging to closed bounded subsets (called the admissible sets) U, V, W, respectively, and $f = [f_1 \ldots f_n]^T$ is a vector-valued function. Such a control system governed by a partial differential equation is called a *distributed parameter system*. Suppose that the boundary-initial conditions for the vector-valued function z are given by $z(a, t) = \phi_1(t)$, $z(b, t) = \phi_2(t)$ and $z(x, t_0) = \psi(x)$, where a, b are constant vectors such that $a \leq x \leq b$ and ϕ_1, ϕ_2 and ψ are known vector-valued functions. The optimal control problem described by the above system and boundary-initial conditions is to find a triple (u^*, v^*, w^*) of control functions such that when all the supplementary constraints imposed on the system as well as all the boundary-initial conditions are satisfied, a given cost functional

$$F\left(z, \frac{\partial z}{\partial x}, \frac{\partial z}{\partial t}, \frac{\partial^2 z}{\partial x \partial x}, \frac{\partial^2 z}{\partial x \partial t}, \frac{\partial^2 z}{\partial t^2}, u, v, w, x, t\right)$$

is minimized, where the terminal time t_1 can be either free or fixed.

Similar to the optimal control theory of systems governed by ordinary differential equations, we also have Pontryagin's maximum principle for certain specific distributed parameter systems. The following simple example is given in Butkouskiy (1969). Consider the system described by

$$\frac{\partial^2 z}{\partial x \partial t} = f\left(z, \frac{\partial z}{\partial x}, \frac{\partial z}{\partial t}, w, x, t\right) \tag{1}$$

where $z \in Z = \mathbb{R}^n$, $t \in [0, t_1]$, and $x \in [0, b]$ with fixed values of t_1 and b. The admissible set W of control functions consists of all such vector-valued functions $w(x, t) = [w_1(x, t) \ldots w_p(x, t)]^T$ where each $w_i(x, t)$ is piecewise continuous and bounded by a function defined on $[0, b] \times [0, t_1]$ with values in some convex closed region in $\mathbb{R}^p$, $p \leq n$. The boundary-initial conditions for the function z is given by $z(0, t) = \phi(t)$ and $z(x, 0) = \psi(x)$. The cost functional to be minimized is given by the Pontryagin function

$$F = c^T z(b, t_1) ,$$

where c is a constant n-vector.

In order to formulate Pontryagin's maximum principle for the above optimal control problem, we introduce the Hamiltonian function

$$H\left(z, \frac{\partial z}{\partial x}, \frac{\partial z}{\partial t}, w, p, x, t\right) = p^T f\left(z, \frac{\partial z}{\partial x}, \frac{\partial z}{\partial t}, w, x, t\right)$$

where $p = [p_1(x, t) \ldots p_n(x, t)]^T$ is determined by

$$\frac{\partial^2 p^T}{\partial x \partial t} = \frac{\partial H}{\partial z} - \frac{d}{dx} \frac{\partial H}{\partial \left(\frac{\partial z}{\partial x}\right)} - \frac{d}{dt} \frac{\partial H}{\partial \left(\frac{\partial z}{\partial t}\right)} ,$$

$$\frac{\partial p^T}{\partial x} = -\frac{\partial H}{\partial \left(\frac{\partial z}{\partial t}\right)} \quad \text{at} \quad t = t_1 , \tag{2}$$

$$\frac{\partial p^T}{\partial t} = -\frac{\partial H}{\partial \left(\frac{\partial z}{\partial x}\right)} \quad \text{at} \quad x = b ,$$

$$p^T(b, t_1) = c^T .$$

Then Pontryagin's maximum principle can be stated as follows: *If $w^*(x, t)$ is an optimal function and $z^*(x, t)$ and $p^*(x, t)$ are the corresponding optimal vector-valued functions defined as above satisfying* (1) *and* (2), *then*

$$H\left(z^*, \frac{\partial z^*}{\partial x}, \frac{\partial z^*}{\partial t}, w^*, p^*, x, t\right) = \max_{w \in W} H\left(z^*, \frac{\partial z^*}{\partial x}, \frac{\partial z^*}{\partial t}, w, p^*, x, t\right)$$

almost everywhere on $[0, b] \times [0, t_1]$.

The optimal control theory of distributed parameter systems is a rapidly developing field. The interested reader is referred to Ahmed and Teo (1981), Butkouskiy (1969, 1983), and Lions (1971).

10.11 Stochastic Optimal Control

Many control systems occurring in practice are affected by certain random disturbances, called noises, which we have ignored in the study of (deterministic) optimal control problems in this book. Stochastic optimal control theory deals with systems in which random disturbances are also taken into consideration. One of the typical stochastic optimal control problems is the linear regulator problem in which the system and observation equations are given by the stochastic differential equations

$$\begin{aligned} d\xi &= [A(t)\xi + B(t)u]\,dt + \Gamma(t)\,dw_1 \\ d\eta &= C(t)\xi\,dt + dw_2, \end{aligned} \qquad t_0 \le t \le t_1 ,$$

and the cost functional to be minimized over an admissible class of control functions is

$$F(u) = E\left\{\int_{t_0}^{t_1} [\xi^T Q(t)\xi + u^T R(t)u]\,dt\right\}.$$

Here the initial state of the system is a Gaussian random vector $\xi(t_0)$, w_1 and w_2 are independent standard Brownian motions with w_2 independent of $\xi(t_0)$, the data vector $\eta(t)$ for $t_0 \le t \le t_1$, t_1 being a fixed terminal time, is known with $\eta(0) = 0$, the matrices $A(t)$, $B(t)$, $C(t)$, $\Gamma(t)$, $Q(t)$ and $R(t)$ are given deterministic matrices of appropriate dimensions with $Q(t)$ being nonnegative definite symmetric and $R(t)$ positive definite symmetric, E is the expectation operator, and the admissible class of control functions consists of Borel measurable functions from $I = [t_0, t_1] \times R^p$ into some closed subet U of I.

Suppose that the control function has partial knowledge of the system states. By this, we mean that the control function u is a linear function of the data rather than the state vector (in the latter case the control function is called a *linear feedback*). For such a linear regulator problem, we have the following *separation principle* which is one of the most useful results in stochastic optimal control theory and shows essentially that the "partially observed" linear regulator problem can be split into two parts: the first is an optimal estimate for the system state using a *Kalman filter*, and the second a "completely observed" linear regulator problem whose solution is given by a linear feedback control function. The separation principle can be stated as follows (Wonham (1968), Fleming and Rishel (1975), Davis (1977), and Kushner (1971)): *An optimal control function for the above partially observed linear regulator problem is given by*

$$u^* = -R^{-1}(t)B^T(t)K(t)\hat{\xi} \ ,$$

where $\hat{\xi}$ is an optimal estimate of ξ from the data $\{\eta: t_0 \leq t \leq t_1\}$, generated by the stochastic differential equation (which induces the standard continuous-time Kalman filter):

$$d\hat{\xi} = [A(t)\hat{\xi} + B(t)u^*] \, dt + H(t)[d\eta - C(t)\hat{\xi} \, dt]$$
$$\hat{\xi}(t_0) = E(\xi(t_0))$$

with $H(t) = P(t)C^T(t)$ and $K(t)$ being the unique solution of the matrix Riccati equation

$$\dot{K}(t) = K(t)B(t)R^{-1}(t)B^T(t)K(t) - K(t)A(t) - A^T(t)K(t) - Q(t), \ t_0 \leq t \leq t_1$$

$$K(t_1) = 0 \ ,$$

and $P(t)$ being the unique solution of the matrix Riccati equation

$$\dot{P}(t) = A(t)P(t) + P(t)A^T(t) + \Gamma(t)\Gamma^T(t) - P(t)C^T(t)C(t)P(t)$$

$$P(t_0) = \text{Var}(\xi(t_0)) \ .$$

The theory of Kalman filtering is an important topic in linear systems and optimal control theory, and as mentioned above, the Kalman filtering process is sometimes needed in stochastic optimal control theory. Discrete-time (or digital) Kalman filter theory and its applications are further investigated in Chui and Chen (1987).

References

Ahmed, H.U., Teo, K.L. (1981): *Optimal Control of Distributed Parameter Systems* (North-Holland, New York)

Balakrishnan, A.V. (1983): *Elements of State Space Theory of Systems* (Optimization Soft-ware, Inc., Publication Division, New York)

Bellman, R.E. (1962): *Applied Dynamic Programming* (Princeton University Press, Princeton, New Jersey)

Boltyanskii, V.G. (1971): *Mathematical Methods of Optimal Control* (Holt, Rinehart and Winston, New York)

Brockett, R.W. (1970): *Finite Dimensional Linear Systems* (John Wiley, New York)

Bryson, A.E., Ho, Y.C. (1969): *Applied Optimal Control* (Blaisdell, Massachusetts)

Butkouskiy, A.G. (1969): *Distributed Control Systems* (American Elsevier Publishing Company, New York)

Butkouskiy, A.G. (1983): *Structural Theory of Distributed Systems* (Ellis Horwood, Chichester)

Casti, J., Kalaba, R. (1983): *Imbedding Methods in Applied Mathematics* (Addison-Wesley, Reading, Massachusetts)

Chen, C.T. (1984): *Linear Systems Theory and Design* (Holt, Rinehart and Winston, New York)

Chen, G., Chui, C.K. (1986): J. Math. Res. Exp., **2**, 75

Chui, C.K., Chen, G. (1987): *Kalman Filtering with Real-Time Applications* (Springer New York)

Chui, C.K., Chen, G.: *Mathematical Approach to Signal Processing and System Theory* (in preparation)

Courant, R., Hilbert, D. (1962): *Methods of Mathematical Physics II* (Interscience, New York)

Davis, M.H.A. (1977): *Linear Estimation and Stochastic Control* (John Wiley, New York)

Fleming, W.H., Rishel, R.W. (1975): *Deterministic and Stochastic Optimal Control* (Springer, New York)

Gamkrelidze, R.V. (1978): *Principles of Optimal Control Theory* (Plenum, New York)

Gilbert, E. (1963): SIAM J. Control, **1**, 128

Hautus, M.L.J. (1973): SIAM J. Control, **11**, 653

Hermann, R. (1984): *Topics in the Geometric Theory of Linear Systems* (Mathematics Sciencs Press, Massachusetts)

Hermes, H., LaSalle, J.P. (1969): *Functional Analysis and Time Optimal Control* (Academic, New York)

Kailath, T. (1980): *Linear Systems* (Prentice-Hall, Englewood Cliffs, New Jersey)

Kalaba, R., Spingarn, K. (1982): *Control, Identification, and Input Optimization* (Plenum, New York)

Kalman, R.E. (1962): Proc. National Acad. Sci., USA, **48**, 596

Kalman, R.E. (1963): SIAM J. Control, **1**, 152

Knowles, G. (1981): *An Introduction of Applied Optimal Control* (Academic, New York)

Kushner, H. (1971): *Introduction to Stochastic Control* (Holt, Rinehart and Winston, New York)

Lee, E.B., Marcus, L. (1967): *Foundations of Optimal Control Theory* (John Wiley, New York)

Lefschetz, S. (1965a): SIAM J. Control, **3**, 1

Lefschetz, S. (1965b): *Stability of Nonlinear Systems* (Academic, New York)

Leigh, J.R. (1983): *Essentials of Nonlinear Control Theory* (Peter Peregrinus, London)

Lewis, F.L. (1986): *Optimal Control* (John Wiley, New York)

Lions, J.L. (1971): *Optimal Control of Systems Governed by Partial Differential Equations* (Springer, New York)

Luenberger, D.G. (1964): IEEE Trans. Military Elec., **3**, 74

Macki, J., Strauss, A. (1982): *Introduction to Optimal Control Theory* (Springer, New York)

Nering, E.D. (1963): *Linear Algebra and Matrix Theory* (John Wiley)

O'Reilly, J. (1983): *Observers for Linear Systems* (Academic, New York)

Padulo, L., Arbib, M.A. (1974): *System Theory* (W.B. Saunders, New York)

Petrov, Iu.P. (1968): *Variational Methods in Optimal Control Theory* (Academic, New York)

Polak, E. (1971): *Computational Methods in Optimization: A Uniform Approach* (Academic, New York)

Pontryagin, L.S., Boltyanskii, V.G., Gamkrelidze, R.V., Mischenko, E.F. (1962): *The Mathematical Theory of Optimal Processes* (John Wiley, New York)

Rolewicz, S. (1987): *Functional Analysis and Control Theory: Linear Systems* (Reidel, Boston)

Royden, H.L. (1968): *Real Analysis* (Macmillan, New York)

Silverman, L.M. (1971): IEEE Trans. Auto. Control, **16**, 554

Sun, C. (1984): Acta Auto. Sinica, **10**, 195

Timothy, L.K., Bona, B.E. (1968): *State Space Analysis* (McGraw-Hill, New York)

Weiss, L. (1969): "Lectures on Controllability and Observability," in *Controllability and Observability* (Centro Int. Matematico, Estivo, Rome) p. 202

Willems, J.C., Miller, S.K. (1971): IEEE Trans. Auto. Control, **16**, 582

Wonham, W.M. (1967): IEEE Trans. Auto. Control, **12**, 660

Wonham, W.M. (1968): SIAM J. Control, **6**, 312

Wonham, W.M. (1974): *Linear Multivariable Systems* (Springer, New York)

Zadeh, L.Z., Desoer, C.A. (1979): *Linear System Theory* (R.E. Kieger, New York)

Answers and Hints to Exercises

Chapter 1

1.1 $\dot{x} = \dfrac{1}{\alpha\delta - \beta\gamma} \begin{bmatrix} a\beta\gamma - b\beta\delta - \alpha\gamma & -a\alpha\beta + b\beta^2 + \alpha^2 \\ a\gamma\delta - b\delta^2 - \gamma^2 & -a\alpha\delta + b\beta\delta + \alpha\gamma \end{bmatrix} x + \begin{bmatrix} \beta \\ \delta \end{bmatrix} u$,

$v = \dfrac{1}{\alpha\delta - \beta\gamma} [\delta \quad -\beta] x$.

1.2 a and b are arbitrary and $c = 0$.

1.3 Since α, β, γ, and δ can be arbitrarily chosen as long as $\alpha\delta - \beta\gamma \neq 0$, the matrices

$$A = \dfrac{1}{\alpha\delta - \beta\gamma} \begin{bmatrix} a\beta\gamma - b\beta\delta - \alpha\gamma & -a\alpha\beta + b\beta^2 + \alpha^2 \\ a\gamma\delta - b\delta^2 - \gamma^2 & -a\alpha\delta + b\beta\delta + \alpha\gamma \end{bmatrix}, \quad B = \begin{bmatrix} \beta \\ \delta \end{bmatrix} \quad \text{and}$$

$$C = \dfrac{1}{\alpha\delta - \beta\gamma} [\delta \quad -\beta] \quad \text{are not unique.}$$

1.4 Let the minimum polynomial of A be $p(\lambda) = p_0\lambda^n + p_1\lambda^{n-1} + \ldots + p_n$ with $p_0 = 1$. Then $a_j = p_j$, $j = 0, 1, \ldots, n$. If $D \neq 0$, then $m = n$ and $b_j = CA^{j-1}B + p_1 CA^{j-2}B + \ldots + p_{j-2}CAB + p_{j-1}CB + p_jD$, $j = 0, 1, \ldots, n$. If $D = 0$, then $m = n-1$ and $b_j = CA^jB + p_1 CA^{j-1}B + \ldots + p_jCB$, $j = 0, 1, \ldots, n-1$.

1.5 (a) Let $x_1 = v_1$, $x_2 = v_1'$, $x_3 = v_2$, $x_4 = v_2'$ and $x = [x_1 \ \ldots \ x_4]^T$. Then

$$\dot{x} = \begin{bmatrix} 0 & 1 & 0 & 0 \\ -a_{12} & -a_{11} & -b_{12} & -b_{11} \\ 0 & 0 & 0 & 1 \\ -a_{22} & -a_{21} & -b_{22} & -b_{21} \end{bmatrix} x + \begin{bmatrix} 0 & 0 \\ \alpha_1 & \beta_1 \\ 0 & 0 \\ \alpha_2 & \beta_2 \end{bmatrix} \begin{bmatrix} u_1 \\ u_2 \end{bmatrix} ,$$

$$\begin{bmatrix} v_1 \\ v_2 \end{bmatrix} = \begin{bmatrix} 1 & 0 & 0 & 0 \\ 0 & 0 & 1 & 0 \end{bmatrix} x .$$

(b)

$$A = \begin{bmatrix} A_{11} & \cdots & A_{1n} \\ \vdots & & \vdots \\ A_{n1} & \cdots & A_{nn} \end{bmatrix}, \quad B = \begin{bmatrix} B_1 \\ \vdots \\ B_n \end{bmatrix},$$

$$C = [C_1, \ldots, C_n] \quad \text{and} \quad D = 0,$$

where

$$A_{ii} = \begin{bmatrix} 0 & 1. & 0 \cdots & 0 \\ & & \ddots & \\ & & & \cdot 1 \\ -a^i_{in} & \cdots & \cdots & -a^i_{i1} \end{bmatrix}_{n \times n},$$

$$A_{ij} = \begin{bmatrix} 0 & \cdots & \cdots & 0 \\ \vdots & & & \vdots \\ 0 & \cdots & \cdots & 0 \\ -a^j_{in} & \cdots & \cdots & -a^j_{i1} \end{bmatrix}_{n \times n},$$

$$j \neq i, \quad i, j = 1, \ldots, n,$$

$$B_i = \begin{bmatrix} 0 & \cdots & 0 \\ \vdots & & \vdots \\ 0 & \cdots & 0 \\ \alpha_{i1} & \cdots & \alpha_{in} \end{bmatrix}_{n \times n}, \quad C_i = \begin{bmatrix} 0 & 0 \ldots 0 \\ \cdots & \\ 1 & 0 \ldots 0 \\ \cdots & \\ 0 & 0 \ldots 0 \end{bmatrix}_{n \times n} \quad \text{(ith row)}$$

$$i = 1, \ldots, n.$$

1.6 (a)

$$x_{k+1} = \begin{bmatrix} 0 & 1 \\ -1 & -1 \end{bmatrix} x_k + \begin{bmatrix} 0 \\ 1 \end{bmatrix} u_k,$$

$$v_k = [1 \quad 0] x_k.$$

(b) Let

$$A = \begin{bmatrix} 0. & 1. & & \\ & \ddots & \ddots & \\ & & 0 & \cdot 1 \\ -a_n & \cdots & \cdots & -a_1 \end{bmatrix}, \quad B = \begin{bmatrix} \beta_1 \\ \vdots \\ \beta_n \end{bmatrix}, \quad C = [1 \ 0 \cdots 0] \text{ and } D = [\beta_0].$$

Then the β_is are determined by

$$
\begin{bmatrix} \beta_n \\ \vdots \\ \beta_1 \\ \beta_0 \end{bmatrix} = \begin{bmatrix} a_0 & a_1 & \cdots & \cdots & a_n \\ & \ddots & \ddots & & \vdots \\ & & \ddots & \ddots & a_1 \\ & & & \ddots & a_0 \end{bmatrix}^{-1} \begin{bmatrix} b_m \\ \vdots \\ b_{m-n+1} \\ b_{m-n} \end{bmatrix} ,
$$

where $a_0 = 1$, $b_j = 0$ for $j < 0$.

Chapter 2

2.1 $\phi(t, t_0) = \begin{bmatrix} e^{t-t_0} & \frac{1}{2}(t^2 - t_0^2)e^{t-t_0} \\ 0 & e^{t-t_0} \end{bmatrix}$.

2.2 $X(\mathcal{U}) = \mathrm{sp}\{1, t, \ldots, t^N, t^{N+1}\}$.

2.3 Let

$$
(t - t_i)_+ = \begin{cases} t - t_i, & \text{if } t \geq t_i , \\ 0, & \text{if } t < t_i . \end{cases}
$$

Then $X(\mathcal{U}) = \mathrm{sp}\{(t - t_0)_+, (t - t_1)_+, \ldots, (t - t_N)_+\}$.

2.4 $X(\mathcal{U}) = \mathrm{sp}\left\{ \begin{bmatrix} 1 \\ 0 \end{bmatrix}, \begin{bmatrix} 0 \\ 1 \end{bmatrix}, \begin{bmatrix} t + t^3/3 \\ t^2/2 \end{bmatrix}, \begin{bmatrix} t^2/2 + t^4/4 \\ t^3/3 \end{bmatrix}, \ldots, \right.$

$\left. \begin{bmatrix} t^{N+1}/(N+1) + t^{N+3}/(N+3) \\ t^{N+2}/(N+2) \end{bmatrix} \right\}$.

2.5 If the input is zero, then the output is $v = Cx = C\Phi(t, t_0)x_0$. Define $v(\cdot) = C\Phi(t, t_0)(\cdot)$. Then $v(ax_{01} + bx_{02}) = av(x_{01}) + bv(x_{02})$. If the initial state is zero, then the output is $v = Cx = C\int_{t_0}^t \Phi(t, s)B(s)u(s)\, ds$. Define $v(\cdot) = C\int_{t_0}^t \Phi(t, s)B(s)(\cdot)\, ds$. Then $v(au_1 + bu_2) = av(u_1) + bv(u_2)$. If (2.10) is considered, then

$$
v_k = C_k A_{k-1} \cdots A_0 x_0 + C_k B_0 u_0 + \cdots + C_k B_{k-1} u_{k-1} + D_k u_k ,
$$

and if (2.11) is considered, then

$$
v(t) = C(t)\Phi(t, t_0)x_0 + C(t)\int_{t_0}^t \Phi(t, s)B(s)u(s)\, ds + D(t)u(t) .
$$

Since $A_0, \ldots, A_{k-1}$ and $\Phi(t, t_0)$ are all nonsingular, the linearity of the output in the input implies that $C_k x_0 = 0$ for all k and $C(t)x_0 = 0$ for all $t \geq t_0$.

2.6   By Hölder's Inequality, we have

$$\int_J |A(t)|_1 \, dt \le \left(\int_J |A(t)|_p^p \right)^{1/p} \left(\int_J 1 \, dt \right)^{1/q} .$$

Suppose that

$$\int_J |A(t)|_p^p \, dt \le C^p < \infty .$$

Then it follows from the Picard iteration process that

$$|P_N(t) - P_M(t)|_1 = \left| \sum_{k=M}^{N-1} \int_{t_0}^{t} A(s_1) \int_{t_0}^{s_1} A(s_2) \ldots \int_{t_0}^{s_k} A(s_{k+1}) \, ds_{k+1} \ldots ds_1 \right|_1$$

$$\le \sum_{k=M}^{N-1} \int_{t_0}^{t} |A(s_1)|_1 \int_{t_0}^{s_1} |A(s_2)|_1 \ldots \int_{t_0}^{s_k} |A(s_{k+1})|_1 \, ds_{k+1} \ldots ds_1$$

$$\le \sum_{k=M}^{N-1} \int_{t_0}^{t} |A(s_1)|_1 \ldots \int_{t_0}^{s_{k-1}} |A(s_k)|_1 \left(\int_{t_0}^{s_k} |A(s_{k+1})|_p^p \right)^{1/p} (s_k - t_0)^{1/q} ds_k \ldots ds_1$$

$$\le \sum_{k=M}^{N-1} C \int_{t_0}^{t} |A(s_1)|_1 \ldots \int_{t_0}^{s_{k-1}} |A(s_k)|_1 (s_k - t_0)^{1/q} \, ds_k \ldots ds_1$$

$$\le \sum_{k=M}^{N-1} C \int_{t_0}^{t} |A(s_1)|_1 \ldots \int_{t_0}^{s_{k-2}} |A(s_{k-1})| \left(\int_{t_0}^{s_{k-1}} |A(s_k)|_p^p (s_k - t_0)^{p/q} \, ds_k \right)^{1/p}$$

$$(s_{k-1} - t_0)^{1/q} \, ds_{k-1} \ldots ds_1$$

$$\le \sum_{k=M}^{N-1} C \int_{t_0}^{t} |A(s_1)|_1 \ldots \int_{t_0}^{s_{k-2}} |A(s_{k-1})| (s_{k-1} - t_0)^{1/q}$$

$$\left(\int_{t_0}^{s_{k-1}} |A(s_k)|_p^p \, ds_k \right)^{1/p} (s_{k-1} - t_0)^{1/q} \, ds_{k-1} \ldots ds_1$$

$$\le \sum_{k=M}^{N-1} C^2 \int_{t_0}^{t} |A(s_1)|_1 \ldots \int_{t_0}^{s_{k-2}} |A(s_{k-1})| (s_{k-1} - t_0)^{2/q} \, ds_{k-1} \ldots ds_1$$

$$\ldots$$

$$\le \sum_{k=M}^{N-1} C^k \int_{t_0}^{t} |A(s_1)|_1 (s_1 - t_0)^{k/q} \, ds_1$$

$$\le \sum_{k=M}^{N-1} C^{k+1} (t - t_0)^{(k+1)/q}$$

which tends to zero uniformly on any bounded interval J as $M, N \to \infty$ independently. The rest of the proof is the same as the proof to the convergence of the infinite series (2.6) in the l_1 norm.

2.7 If the discretization formulas $A_k = hA(kh) + I$ etc. are used, then only the vector $[a \;\; b]^T = [-11/15 \;\; -11/25]^T$ can be brought to the origin in two steps when $h = 1/5$ and only the vector $[a \;\; b]^T$ satisfying $1210a - 550b + 336 = 0$ can be brought to the origin in two steps when $h = 1/10$. If the discretization formulas $\Phi_{ij} = \Phi(ih, jh)$ etc. are used, then only the vector

$$
\begin{bmatrix} a \\ b \end{bmatrix} =
$$

$$
- \begin{bmatrix} e^h - 1 & -\dfrac{1}{2} - \dfrac{5}{11}e^h - \dfrac{1}{22}e^{-10h} \\[2ex] 0 & \dfrac{1}{10}(1 - e^{-10h}) \end{bmatrix}^{-1}
\begin{bmatrix} -\dfrac{1}{2} + \dfrac{16}{11}e^h + \dfrac{1}{22}e^{-10h} \\[2ex] \dfrac{1}{10}(e^{-10h} - 1) \end{bmatrix} (h + 2)
$$

can be brought to the origin in two steps.

2.8 By Hölder's Inequality, we have

$$\sum_{i,j} |a_{ij} + b_{ij}|^{p-1} |a_{ij}| \le \left(\sum_{i,j} |a_{ij} + b_{ij}|^{q(p-1)} \right)^{1/q} \left(\sum_{i,j} |a_{ij}|^p \right)^{1/p}$$

$$= \left(\sum_{i,j} |a_{ij} + b_{ij}|^p \right)^{1/q} \left(\sum_{i,j} |a_{ij}|^p \right)^{1/p} \qquad \text{and}$$

$$\sum_{i,j} |a_{ij} + b_{ij}|^{p-1} |b_{ij}| \le \left(\sum_{i,j} |a_{ij} + b_{ij}|^{q(p-1)} \right)^{1/q} \left(\sum_{i,j} |b_{ij}|^p \right)^{1/p}$$

$$= \left(\sum_{i,j} |a_{ij} + b_{ij}|^p \right)^{1/q} \left(\sum_{i,j} |b_{ij}|^p \right)^{1/p}.$$

Hence, we have

$$|A + B|_p^p = \sum_{i,j} |a_{ij} + b_{ij}|^p \le \sum_{i,j} |a_{ij} + b_{ij}|^{p-1} |a_{ij}| + \sum_{i,j} |a_{ij} + b_{ij}|^{p-1} |b_{ij}|$$

$$\le \left(\sum_{i,j} |a_{ij} + b_{ij}|^p \right)^{1/q} \left[\left(\sum_{i,j} |a_{ij}|^p \right)^{1/p} + \left(\sum_{i,j} |b_{ij}|^p \right)^{1/p} \right]$$

so that

$$\left(\sum_{i,j} |a_{ij}+b_{ij}|^p\right)^{1-1/q} \le \left(\sum_{i,j} |a_{ij}|^p\right)^{1/p} + \left(\sum_{i,j} |b_{ij}|^p\right)^{1/p},$$

i.e., $|A+B|_p \le |A|_p + |B|_p$.

Chapter 3

3.1 If $x_1, x_2 \in V_t$, then there exist two controls u_1 and u_2 such that $0 = \Phi(t, t_0)x_i + \int_{t_0}^t \Phi(t,s)B(s)u_i(s)\,ds$, $i=1, 2$. Thus, $0 = \Phi(t,t_0)(ax_1 + bx_2) + \int_{t_0}^t \Phi(t,s)B(s)(au_1(s) + bu_2(s))\,ds$; i.e., $(ax_1 + bx_2) \in V_t$. If x_0 can be brought to 0 at time s by a control u, then it can also be brought to 0 at time $t \ge s$ by

$$\tilde{u}(\tau) = \begin{cases} u(\tau), & \text{if } t_0 \le \tau \le s \\ 0, & \text{if } s < \tau \le t. \end{cases}$$

Hence, V_s is a subspace of V_t if and only if $s \le t$. Combining the above two facts, we can similarly prove that V is a subspace of $\mathbb{R}^n$.

3.2 Let $x = x_1 + x_2$ where $x_1 \in (vR)^\perp$ and $x_2 \in vR$. If $y \in \text{Im}\{R\}$, then there is a z such that $y = Rz$ and so $y^T x_2 = z^T R^T x_2 = z^T R x_2 = 0$. Hence, $y \in (vR)^\perp$, i.e., $\text{Im}\{R\} \subseteq (vR)^\perp$. By linear algebra, $\dim(\text{Im}\{R\}) = \dim(vR)^\perp$. Hence, $\text{Im}\{R\} = (vR)^\perp$. Suppose that $x = 0$. If $x_1 \ne 0$, then $0 = x_1^T x = x_1^T(x_1 + x_2) = x_1^T x_1 \ne 0$, a contradiction. If $x_2 \ne 0$, we have the same contradiction. Hence $x = x_1 + x_2 = 0$.

3.3 If $\mathscr{S}$ is controllable, then for any x_0, there is a u such that

$$\int_{t_0}^{t^*} \Phi(t^*, s)B(s)u(s)\,ds = -\Phi(t^*, t_0)x_0 ;$$

i.e., $\text{Im}\{L_{t^*}\} = \mathbb{R}^n$. By Lemma 3.2, $\text{Im}\{Q_{t^*}\} = \mathbb{R}^n$. Hence, Q_{t^*} is nonsingular. If Q_{t^*} is nonsingular, let $u(s) = B^T(s)\Phi^T(t^*, s)y$. Then for any x_0 the equation

$$\Phi(t^*, t_0)x_0 + \left[\int_{t_0}^{t^*} \Phi(t^*, s)B(s)B^T(s)\Phi^T(t^*, s)\,ds\right]y = 0$$

has a unique solution y. Hence, $\mathscr{S}$ is controllable.

3.4 $\det Q_t = \frac{1}{12}(t-t_0)^4 \ne 0$ for all $t > t_0$.

3.5 $\det Q_t = (b^4/12)(t-t_0)^4 \ne 0$ for all $t > t_0$ and $b \ne 0$.

3.6 Verify that $\Phi(t^*, t_0)y_0 + \int_{t_0}^{t^*} \Phi(t^*, s)B(s)u^*(s)\,ds = y_1$.

3.7 By the Cayley-Hamilton Theorem, $A^m = 0$ for $m \ge n$. Hence

$$a^T e^{bA} = a^T\left(I + bA + \ldots + \frac{b^n}{n!}A^n + \ldots\right) = 0 .$$

3.8 Since $\Phi_{n0} = 0$ for all $n \geq 1$, even the zero control sequence can bring any y_0 to the origin. But since the last rows of A_k and B_k are all zero, the last row on the left-hand side of $\Phi_{n0}y_0 + \Sigma_{k=1}^n \Phi_{nk}B_{k-1}u_{k-1} = y_1$ is always zero. Hence no control sequence can bring $y_0 = 0$ to $y_1 = [0 \ldots 0 \quad 1]^T$. Since

$$x_k = \begin{bmatrix} x_{k1} \\ x_{k2} \end{bmatrix} = \begin{bmatrix} 10^k a - 10^{k-1}u_{01} - 10^{k-2}u_{11} - \ldots - u_{k-1,1} \\ -10^{k-1}a + 10^{k-2}u_{01} + 10^{k-3}u_{11} + \ldots + 0.1u_{k-1,1} \end{bmatrix}$$

$$= \begin{bmatrix} -10x_{k2} \\ x_{k2} \end{bmatrix} = \begin{bmatrix} -10 \\ 1 \end{bmatrix} x_{k2} \, ,$$

any control sequence which brought x_{k2} to 0 will bring x_{k1} to 0. But any control sequence which brought x_{k2} to 0 cannot bring x_{k1} to 1, i.e., $\begin{bmatrix} a \\ b \end{bmatrix}$ cannot be brought to $\begin{bmatrix} 1 \\ 0 \end{bmatrix}$.

3.9 For any given initial state $x_0 = \begin{bmatrix} a \\ b \end{bmatrix}$, we always have $x_2 = \begin{bmatrix} u_0 \\ u_1 \end{bmatrix}$. Hence, $\begin{bmatrix} a \\ b \end{bmatrix}$ can be brought to any preassigned position $\begin{bmatrix} y_1 \\ y_2 \end{bmatrix}$ provided that the control is chosen to be $\begin{bmatrix} u_0 \\ u_1 \end{bmatrix} = \begin{bmatrix} y_1 \\ y_2 \end{bmatrix}$.

3.10 If R_{l^*} is a nonsingular matrix and $u_{i-1} = B_{i-1}^T \Phi_{l^*i}^T z$, then $\Phi_{l^*l}y_0 + (\Sigma_{i=l+1}^{l^*} \Phi_{l^*i}B_{i-1}B_{i-1}^T \Phi_{l^*i}^T)z = y$ has a unique solution z; i.e., $\mathscr{S}$ is controllable. If $\mathscr{S}$ is controllable, then for any x_0, there is $\{u_i\}$ such that $\Phi_{l^*l}x_0 + \Sigma_{i=l+1}^{l^*} \Phi_{l^*i}B_{i-1}u_{i-1} = 0$; i.e., $y_0 = -\Phi_{l^*l}x_0$ is in the image of R_{l^*}. Since x_0 is arbitrary, $\mathrm{Im}\{R_{l^*}\} = \mathbb{R}^n$; i.e., R_{l^*} is nonsingular.

3.11 If $\mathscr{S}$ is controllable, then by Theorem 3.6, R_{l^*} is nonsingular. The universal control sequence $u_k^* = B_k^T \Phi_{l^*k+1}^T R_{l^*}^{-1}(y_1 - \Phi_{l^*l}y_0)$ then satisfies $\Phi_{l^*l}y_0 + \Sigma_{i=l+1}^{l^*} \Phi_{l^*i}B_{i-1}u_{i-1} = y_1$; i.e., $\mathscr{S}$ is completely controllable.

3.12 $\mathscr{S}$ is (completely) controllable if and only if the matrix M_{AB} has rank n, and this is equivalent to saying that (3.14) has a solution $u_1, \ldots, u_{n+l-1}$; i.e., a universal discrete time-interval can be chosen such that its "length" is n. Consider the example

$$x_{k+1} = \begin{bmatrix} 1 & 1 \\ 0 & 1 \end{bmatrix} x_k + \begin{bmatrix} 0 \\ 1 \end{bmatrix} u_k \, , \qquad x_0 = \begin{bmatrix} a \\ b \end{bmatrix} .$$

3.13 $\det M_{AB} = acd + bc^2 - d^2 \neq 0$.
3.14 $\det M_{AB} = ac - b - c^2 \neq 0$.

Chapter 4

4.1 For any $t_0 \geq 0$, there exists a $t_1 > \max(t_0, 1)$ such that

$$\begin{bmatrix} v(t_0) \\ v(t_1) \end{bmatrix} = \begin{bmatrix} 1 & (1-t_0) - |t_0 - 1| \\ 1 & 2(1-t_1) \end{bmatrix} \begin{bmatrix} x_{01} \\ x_{02} \end{bmatrix} ,$$

where the coefficient matrix is always nonsingular. But if $0 \le t_0 < 1$, then the coefficient matrix becomes $\begin{bmatrix} 1 & 0 \\ 1 & 0 \end{bmatrix}$ on $(t_0, t_1) \subset (t_0, 1)$.

4.2 The corresponding coefficient matrix is

$$\begin{bmatrix} 1 & 2(1-t_0) \\ 1 & 1-t_1+|t_1-1| \end{bmatrix}$$

which is nonsingular for any $t_0 \in [0, 1)$ and $t_1 > t_0$. But the matrix becomes $\begin{bmatrix} 1 & 0 \\ 1 & 0 \end{bmatrix}$ for any $t_0 \ge 1$ and $t_1 > t_0$.

4.3 a and b are arbitrary.

4.4 $\det N_{CA} = -b^2$; $\det P_t = b^3[(t-t_0)/12 - a](t-t_0)^2$ which is nonzero for some $t > t_0$ if and only if $b \ne 0$; a can be arbitrary.

4.5 $\mathscr{S}$ has the observability property on $\{l, \ldots, m\}$ if and only if

$$\begin{bmatrix} C_l \\ C_{l+1}\Phi_{l+1,l} \\ \vdots \\ C_m\Phi_{ml} \end{bmatrix} x_l = \begin{bmatrix} v_l \\ v_{l+1} \\ \vdots \\ v_m \end{bmatrix}$$

has a unique solution x_l, and this is equivalent to the coefficient matrix being of full (column) rank, or $x_l = 0$ whenever (4.6) holds for $k = l, \ldots, m$.

4.6 Suppose that $\mathscr{S}$ is observable at time l. Then there is a $p > l$ such that x_l is uniquely determined by $(0, v_k)$, $k = l, \ldots, p$. If L_m is singular for all $m > l$, then $y_l^T L_p y_l = 0$ for some $y_l \ne 0$, i.e., $C_k\Phi_{kl}y_l = 0$, $k = l, \ldots, p$. But for $u_k = 0$ we have $v_k = C_k\Phi_{kl}x_l$, $k = l, \ldots, p$, so that $v_k = C_k\Phi_{kl}(x_l + \alpha y_l)$ for $k = l, \ldots, p$ and arbitrary α, a contradiction. Suppose that L_p is nonsingular for some $p > l$. Then it can be shown, by using (3.9) and (4.5), that

$$L_p x_l = \sum_{k=l+1}^{p} \Phi_{kl}^T C_k^T v_k - \sum_{k=l+1}^{p} \Phi_{kl}^T C_k^T D_k u_k$$

$$- \sum_{k=l+1}^{p} \sum_{i=l+1}^{k} \Phi_{kl}^T C_k^T C_k \Phi_{ki} B_{i-1} u_{i-1} \, ,$$

so that x_l is uniquely determined by u_k and v_k over $\{l, \ldots, p\}$.

4.7 If the rank of N_{CA} is less than n, then there is an $a \ne 0$ such that $Ca = CAa = \ldots = CA^{n-1}a = 0$. By the Cayley-Hamilton Theorem, $CA^{k-1}a = 0$ for all $k \ge l$ so that $L_m a = 0$ for all $m > l$. Hence L_m is singular for all $m > l$ so that, by Theorem 4.3, $\mathscr{S}$ is not observable at time l.

Suppose that N_{CA} has rank n. Let x_l and y_l be two initial states determined by the same (u_k, v_k), $k = l, \ldots, m$. Then it is easy to obtain $N_{CA}(x_l - y_l) = 0$, so that $x_l = y_l$; i.e., $\mathscr{S}$ is observable at time l.

4.8 Suppose that $\mathscr{S}$ is totally observable. Then $C_k A^{k-l} x_l = 0$ for $k = l$, $l+1$. Hence $x_l = 0$. It implies that $T_{CA} = \begin{bmatrix} C \\ CA \end{bmatrix}$ has rank n. Conversely, if T_{CA} has rank n, then whenever $C_k A^{k-l} x_l = 0$ for $k = l$, $l+1$, we must have $x_l = 0$. Hence $\mathscr{S}$ is totally observable.

4.9 Let $\Phi(t,s)$ and $\Psi(t,s)$ be the transition matrices of $A(t)$ and $-A^T(t)$, respectively. Then $\mathscr{S}$ is controllable on $(t_0, t^*) \Leftrightarrow Q_{t^*}$ is nonsingular $\Leftrightarrow \int_{t_0}^{t^*} \Phi(t_0, t) B(t) B^T(t) \Phi^T(t_0, t) dt$ is nonsingular, or equivalently $P_{t^*} = \int_{t_0}^{t^*} \Psi^T(t, t_0) B(t) B^T(t) \Psi(t, t_0) dt$ is nonsingular (Lemma 4.1) $\Leftrightarrow \mathscr{S}$ has the observability property on (t_0, t^*). Conversely, $\mathscr{S}$ has the observability property on $(t_0, t_1) \Leftrightarrow P_{t_1}$ is nonsingular $\Leftrightarrow \int_{t_0}^{t_1} \Psi(t_0, t) C^T(t) C(t) \Psi^T(t_0, t) dt$ is nonsingular (Lemma 4.1) $\Leftrightarrow Q$ is nonsingular $\Leftrightarrow \mathscr{S}$ is controllable on (t_0, t_1).

4.10 $\mathscr{S}_d$ is completely controllable with the universal discrete time-interval $\{l, \ldots, l^*\}$ if and only if the matrix $R_{l^*} = \Sigma_{i=l+1}^{l^*} \Phi_{l^* i} B_{i-1} B_{i-1}^T \Phi_{l^* i}^T = \Sigma_{i=l+1}^{l^*} (A_{l^*-1} \ldots A_i) B_{i-1} B_{i-1}^T (A_i^T \ldots A_{l^*-1}^T)$ is nonsingular. Multiplying both sides to the left by $(A_{l^*-1} \ldots A_l)^{-1}$ and to the right by $(A_l^T \ldots A_{l^*-1}^T)^{-1}$, it is equivalent to the nonsingularity of the observability matrix L_{l^*} of the system $\mathscr{S}_d$ where $L_{l^*} = \Sigma_{i=l+1}^{l^*} \Psi_{il} B_{i-1} B_{i-1}^T \Psi_{il} + \Sigma_{i=l+1}^{l^*} [(A_{i-1}^{-1})^T \ldots (A_l^{-1})^T]^T B_{i-1} B_{i-1}^T [(A_{i-1}^{-1})^T \ldots (A_l^{-1})^T]$. Finally, L_{l^*} is nonsingular if and only if $\mathscr{S}_d$ has the observability property on $\{l, \ldots, l^*\}$. Similarly, $\mathscr{S}_d$ has the observability property on $\{l, \ldots, m\}$ if and only if $L_m = \Sigma_{i=l+1}^{m} (A_{i-1} \ldots A_l)^T C_i^T C_i (A_{i-1} \ldots A_l)$ is nonsingular. Multiplying to the left by $[(A_{l^*-1}^{-1})^T \ldots (A_l^{-1})^T]$ and to the right by $(A_l^{-1} \ldots A_{l^*-1}^{-1})$, it is equivalent to the nonsingularity of the controllability matrix R_m of $\mathscr{S}_d$, which is equivalent to $\mathscr{S}_d$ being controllable with $\{l, \ldots, m\}$ as a universal discrete time-interval.

4.11 If $c = 0$ and $a \neq 0$, or if $c \neq 0$ and a and b are arbitrary, then $\mathscr{S}$ is completely observable. If $c \neq 0$, $\mathscr{S}$ is always totally observable; otherwise it is always not observable.

4.12 For all a and b, $\mathscr{S}$ is always completely observable. The input-output relation of its dual system is $\tilde{v}_{k+3} + (a-1)\tilde{v}_{k+2} - \tilde{v}_{k+1} = \tilde{u}$. The dual system is completely observable if and only if $a \neq 1$.

4.13 $r(N_{CA}) =$

$$r\left(\begin{bmatrix} C \\ CA \\ \vdots \\ CA^{n-1} \end{bmatrix} (A^{-1})^{n-1}\right) = r\left(\begin{bmatrix} C(A^{-1})^{n-1} \\ C(A^{-1})^{n-2} \\ \vdots \\ C \end{bmatrix}\right)$$

$$= r\left(\begin{bmatrix} C \\ CA^{-1} \\ \vdots \\ C(A^{-1})^{n-1} \end{bmatrix}\right) = r(N_{CA^{-1}}).$$

Chapter 5

5.1 Since $M_{\tilde{A}\tilde{B}} = G^{-1}M_{AB}$ and $N_{\tilde{C}\tilde{A}} = N_{CA}G$, the nonsingular transformation G does not change the ranks of M_{AB} and N_{CA}. Since the transition matrix of the transformed system is $\tilde{\Phi}(t, s) = G^{-1}\Phi(t, s)G$, $\tilde{Q}_{t^*} = G^{-1}Q_{t^*}(G^{-1})^T$ and $\tilde{P}_t = G^T P_t G$.

5.2 If the system $\mathscr{S}$ with zero transfer matrix D is completely controllable, then Q_{t^*} is nonsingular. Hence, a universal time-interval $(t_0, t^*) \subset J$ and a universal control u^* exist for the same system with a nonzero transfer matrix D such that the equation

$$\Phi(t^*, t_0)y_0 + \int_{t_0}^{t^*} \Phi(t^*, s)B(s)u(s)\,ds = y_1$$

has an admissible solution u^* for arbitrarily given y_0 and y_1.

If the system $\mathscr{S}$ with zero transfer matrix is observable at time t_0, then there exists an interval $(t_0, t_1) \subset J$ such that $(u(t), v(t))$, $t_0 \le t \le t_1$, uniquely determines an initial state $x(t_0)$. Hence, it can be shown that the equation

$$C(t)\Phi(t, t_0)x(t_0) = v(t) - D(t)u(t) + \int_{t_0}^{t} C(t)\Phi(t, s)B(s)u(s)\,ds$$

has a unique solution $x(t_0)$ for an arbitrarily given pair $(u(t), v(t))$.

5.3 $x(t) = \Phi(t, t_0)x_0 + \int_{t_0}^{t} \Phi(t, s)[B(s)u(s) + f(s)]\,ds$

$$= \Phi(t, t_0)[x_0 + \int_{t_0}^{t} \Phi(t_0, s)f(s)\,ds] + \int_{t_0}^{t} \Phi(t, s)B(s)u(s)\,ds$$

$$:= \Phi(t, t_0)y_0 + \int_{t_0}^{t} \Phi(t, s)B(s)u(s)\,ds$$

$$C(t)\Phi(t, t_0)x(t_0) = v(t) - D(t)u(t) + \int_{t_0}^{t} C(t)\Phi(t, \tau)[B(\tau)u(\tau) + f(\tau)]\,d\tau$$

$$= [v(t) + \int_{t_0}^{t} C(t)\Phi(t, \tau)f(\tau)\,d\tau] - D(t)u(t) + \int_{t_0}^{t} C(t)\Phi(t, \tau)B(\tau)u(\tau)\,d\tau$$

$$:= v_0(t) - D(t)u(t) + \int_{t_0}^{t} C(t)\Phi(t, \tau)B(\tau)u(\tau)\,d\tau \ .$$

5.4 Consider the linear system $\mathscr{S}$ with discrete-time state-space description

$$x_{k+1} = A_k x_k + B_k u_k$$

$$v_k = C_k x_k + D_k u_k \ .$$

If $\{G_k\}$ is any sequence of nonsingular constant matrices and the state vector x_k is changed to y_k by $y_k = G_k^{-1} x_k$, then the matrices A_k, B_k, C_k, and D_k are automatically changed to $\tilde{A}_k = G_k^{-1} A_k G_k$, $\tilde{B}_k = G_k^{-1} B_k$, $\tilde{C}_k = C_k G_k$ and $\tilde{D}_k = G_k^{-1} D_k$ respectively. Hence, the transition matrix of $\mathscr{S}$ is changed from $\Phi_{kj} = A_{k-1} \ldots A_j$ to $\tilde{\Phi}_{kj} = G_{k-1}^{-1} A_{k-1} G_{k-1} G_{k-2}^{-1} A_{k-2} G_{k-2} \cdots G_j^{-1} A_j G_j$ and the matrices R_{l^*} and L_m are changed to

$$\tilde{R}_{l^*} = \sum_{i=\bar{l}+1}^{l^*} \tilde{\Phi}_{l^*i} G_{i-1}^{-1} B_{i-1} B_{i-1}^T [G_{i-1}^{-1}]^T \tilde{\Phi}_{l^*i}^T \quad \text{and}$$

$$\tilde{L}_m = \sum_{k=\bar{l}+1}^{m} \tilde{\Phi}_{kl}^T G_k^T C_k^T C_k G_k \tilde{\Phi}_{kl} \; ,$$

respectively. Moreover, $\tilde{R}_{l^*}$ and $\tilde{L}_m$ have the same ranks as R_{l^*} and L_m, respectively.

The transfer matrices D_k can be assumed to be zero in the study of controllability and observability. The control equation can be extended to include a sequence of vector-valued functions, i.e., $x_{k+1} = A_k x_k + B_k u_k + f_k$ without the controllability and observability properties being changed.

The justification of the above statements is similar to the answers to the previous three exercises.

5.5 Let x be in V_4. Then $x \in \mathrm{sp}\{N_{CA}^T\}$ so that $A^T x \in \mathrm{sp}\{N_{CA}^T\} = V_2 \oplus V_4$. Hence $A^T x = x_2 + x_4$ where $x_2 \in V_2$ and $x_4 \in V_4$. Since $Ax_2 \in \mathrm{sp}\{M_{AB}\}$ which is orthogonal to V_4, we have

$$x_2^T x_2 = (x_2 + x_4)^T x_2 = (A^T x)^T x_2 = x^T A x_2 = 0 \; .$$

Hence, $x_2 = 0$ and $A^T x = x_4 \in V_4$.

5.6 Let $W = [w_{ij}]_{4 \times 4}$ and $\tilde{A} = W^T A W = [\tilde{a}_{ij}]_{4 \times 4}$ with $\tilde{a}_{ij} = 0$ if $i > j$. Then, since W is a unitary matrix, we have $W\tilde{A} = AW$. Comparing the $(1, 1)$ entry and the $(2, 1)$ entry, we have $w_{11} \tilde{a}_{11} = w_{11} + w_{21}$ and $w_{21} \tilde{a}_{11} = w_{21}$, respectively, so that $w_{21} = 0$. Thus, $\tilde{B} = W^T B = [0 \; w_{22} \; w_{23} \; w_{24}]^T$, and $\mathscr{S}_1$ is not controllable.

5.7 Use the definitions of M_{AB} and N_{CA} directly.

5.8 Since any nonsingular transformation does not change the ranks of M_{AB} and N_{CA} (Exercise 5.1), the dimensions of V_1, V_2, V_3 and V_4 are never changed.

5.9 Since

$$U^{-1} M_{AB} = [\tilde{B} \quad \tilde{A}\tilde{B} \ldots \tilde{A}^{n-1}\tilde{B}] =$$

$$\begin{bmatrix} \begin{bmatrix} B_1 \\ B_2 \end{bmatrix} & \begin{bmatrix} A_{11} & A_{12} \\ 0 & A_{22} \end{bmatrix}\begin{bmatrix} B_1 \\ B_2 \end{bmatrix} & \cdots & \begin{bmatrix} A_{11} & A_{12} \\ 0 & A_{22} \end{bmatrix}^{n-1}\begin{bmatrix} B_1 \\ B_2 \end{bmatrix} \\ 0 & 0 & \cdots & 0 \\ 0 & 0 & \cdots & 0 \end{bmatrix}$$

and rank $(U^{-1}M_{AB}) = \text{rank}(M_{AB}) = n_1 + n_2$, we have

$$\text{rank}\left(\left[\begin{bmatrix} B_1 \\ B_2 \end{bmatrix}\begin{bmatrix} A_{11} & A_{12} \\ 0 & A_{22} \end{bmatrix}\begin{bmatrix} B_1 \\ B_2 \end{bmatrix} \cdots \begin{bmatrix} A_{11} & A_{12} \\ 0 & A_{22} \end{bmatrix}^{n-1}\begin{bmatrix} B_1 \\ B_2 \end{bmatrix}\right]\right) = n_1 + n_2;$$

i.e., the combined subsystem $\mathscr{S}_1$ and $\mathscr{S}_2$ is (completely) controllable. Since the above shows that

$$\text{rank}\left(\begin{bmatrix} B_1 & * & \cdots & * \\ B_2 & A_{22}B_2 & \cdots & A_{22}^{n-1}B_2 \end{bmatrix}\right) = n_1 + n_2$$

where the $*$ entries are in terms of A_{11}, A_{12}, A_{22}, B_1 and B_2, we have

$$\text{rank}([B_2 \quad A_{22}B_2 \cdots A_{22}^{n-1}B_2]) = n_2 \ ,$$

so that $\mathscr{S}_2$ is also (completely) controllable. (*Note*: this does not imply that $\mathscr{S}_1$ is also (completely) controllable because the rank of $[B_1 \quad A_{11}B_1 \cdots A_{11}^{n-1}B_1]$ may not be n_1, see (5.3) and Exercise 5.13b). The observability can be similarly proved.

5.10 $Z\{g_{k+1}\} = \sum_{k=0}^{\infty} g_{k+1}z^{-k} = -zg_0 + z\sum_{k=0}^{\infty} g_k z^{-k} = -zg_0 + zZ\{g_k\} \ .$

$Z\{g_{k+j}\} = \sum_{k=0}^{\infty} g_{k+j}z^{-k} = -z^j(g_0 + g_1 z^{-1} + \ \cdots \ + g_{j-1}z^{-(j-1)})$

$+ z^j(g_0 + g_1 z^{-1} + \ \cdots)$

$= -z^j \sum_{i=0}^{j-1} g_i z^{-i} + z^j Z\{g_k\} \ .$

5.11 $H(s) = \dfrac{(s-1)}{(s+3)(s-1)} = \dfrac{1}{s+3}$, $r(M_{AB}) = 1$ and $r(N_{CA}) = 2$.

5.12 $q_m(s) - q_m(t) = (s^m - t^m) - a_1(s^{m-1} - t^{m-1}) - \ \cdots \ - a_{m-1}(s-t)$

$= (s-t)[(s^{m-1} + s^{m-2}t + \ \cdots \ + st^{m-2} + t^{m-1})$

$- a_1(s^{m-2} + s^{m-3} + \ \cdots \ + st^{m-3} + t^{m-2}) - \ \cdots \ - a_{m-1}]$

$= (s-t)[s^{m-1} + s^{m-2}(t-a_1) + s^{m-3}(t^2 - a_1 t - a_2) + \ \cdots$

$+ s(t^{m-2} - a_1 t^{m-3} - \ \cdots \ - a_{m-2}t)$

$+ (t^{m-1} - a_1 t^{m-2} - \ \cdots \ - a_{m-1})]$

$= (s-t)\sum_{k=0}^{m-1}(t^k - a_1 t^{k-1} - \ \cdots \ - a_k)s^{m-k-1} \ .$

5.13 Use the definitions of M_{AB} and N_{CA} directly.

Chapter 6

6.1 (a)

$$I - \Phi(t, t_0) = \begin{bmatrix} 0 & 0 & -(t - t_0) \\ 0 & 0 & 0 \\ 0 & 0 & 0 \end{bmatrix},$$

$$\{\text{equilibrium points}\} = \text{sp} \left\{ \begin{bmatrix} 1 \\ 0 \\ 0 \end{bmatrix}, \begin{bmatrix} 0 \\ 1 \\ 0 \end{bmatrix} \right\}.$$

(b)

$$I - \Phi(t, t_0) = \begin{bmatrix} 0 & 0 & 0 \\ 0 & 0 & 0 \\ 0 & 0 & -(t, t_0) \end{bmatrix},$$

$$\{\text{equilibrium points}\} = \text{sp} \left\{ \begin{bmatrix} 1 \\ 0 \\ 0 \end{bmatrix}, \begin{bmatrix} 0 \\ 1 \\ 0 \end{bmatrix} \right\}.$$

6.2 (a)

$$I - \Phi(t, t_0) = \begin{bmatrix} 0 & -\frac{1}{2}(t^2 - t_0^2) \\ 0 & 0 \end{bmatrix}, \quad \{\text{equilibrium points}\} = \text{sp} \left\{ \begin{bmatrix} 1 \\ 0 \end{bmatrix} \right\}.$$

(b) $I - \Phi(t, t_0) = \begin{bmatrix} 1 - \cosh(t - t_0) & 1 - \sinh(t - t_0) \\ 0 & 0 \end{bmatrix},$

$$\{\text{equilibrium points}\} = \text{sp} \left\{ \begin{bmatrix} 1 \\ \dfrac{\cosh(t - t_0) - 1}{1 - \sinh(t - t_0)} \end{bmatrix} \right\}.$$

6.3   Since $A(t)x_e = \dot{x}_e = 0$ and $A(t)$ is nonsingular at some $t > t_0$, we have $x_e = 0$.

6.4   Let $E = [e_{ij}]$ and $F = [f_{ij}]$. Then

$$|EF|_2^2 = \sum_{i,k} \left(\sum_j e_{ij} f_{jk} \right)^2 \leq \sum_{i,k} \left(\sum_j e_{ij}^2 \sum_j f_{jk}^2 \right) = \sum_{i,j} e_{ij}^2 \sum_{j,k} f_{jk}^2 = |E|_2 |F|_2 .$$

6.5 $|A|_p = |(A + B) - B|_p \leq |A + B|_p + |B|_p$ implies that $|A|_p - |B|_p \leq$

$|A + B|_p$, and $|B|_p = |(A + B) - A|_p \leq |A + B|_p + |A|_p$ implies that $|B|_p -$

$|A|_p \leq |A + B|_p$. Hence $| \, |A|_p - |B|_p| \leq |A + B|_p$.

6.6 $\left| \int_a^b F(t)\,dt \right|_p = \left| \lim_{n\to\infty} \sum_{i=1}^n F(t_i)\,\Delta t_i \right|_p \le \lim_{n\to\infty} \sum_{i=1}^n |F(t_i)|_p \Delta t_i = \int_a^b |F(t)|_p\,dt$.

6.7 Suppose not. Then there is some entry $\phi_{i_0,\,j_0}(t, t_0)$ in $\Phi(t, t_0)$, $1 \le i_0$, $j_0 \le n$, such that $|\phi_{i_0,\,j_0}(t_M, t_0)| > \varepsilon_0$ for $t_M > t_0$ and some $\varepsilon_0 > 0$. Let $x(t_0) = [0 \ldots 0\ 1\ 0 \ldots 0]^T = e_{j_0}$. Then

$$|x(t_M)| = |\Phi(t_M, t_0)x(t_0)| \ge |\phi_{i_0,\,j_0}(t_M, t_0)| > \varepsilon_0\ ;$$

i.e., $|x(t)| \not\to 0$ as $t \to +\infty$, contradicting the asymptotical stability assumption.

6.8 (a) $\lim_{t\to\infty} e^{-at}t^b = \lim_{t\to\infty} (t^b/e^{at})$. Use L'Hospital's rule.
(b) Without loss of generality, suppose that $c > 0$. Write $c = \exp(\ln c)$. Since $c < 1$, $\ln c < 0$. Hence, from (a) we have

$$\lim_{m\to\infty} m^a c^m = \lim_{m\to\infty} e^{(\ln c)m} m^a = 0\ .$$

6.9 Let c satisfy $0 < c < a$. Then for large values of t, $ct \le (a-b)t - \ln M$. Hence,

$$|f(t)| \le M e^{-at}\, t^b = e^{\ln M}\, e^{-at}\, e^{b\ln t} \le e^{-[(a-b)t - \ln M]} \le e^{-ct}$$

for all large values of t.

6.10 The time-invariant free system (6.1) is asymptotically stable about 0 if and only if $|\Phi(t, t_0)| \to 0$ as $t \to +\infty$ by Theorem 6.1, where $\Phi(t, t_0)$ is given by (6.6), if and only if $\mathrm{Re}\{\lambda_j\} < 0$ for all j. Similarly, the system is stable about 0 if and only if there exists some constant $C > 0$ such that $|\Phi(t, t_0)| \le C$ by Theorem 6.1, and this is equivalent to $\mathrm{Re}\{\lambda_j\} < 0$ for all j and λ_j is a simple eigenvalue of A whenever $\mathrm{Re}\{\lambda_j\} = 0$. This statement can be concluded by examining (6.6).

6.11 Denote

$$E = \begin{bmatrix} 0 & 1 & & \\ & \ddots & \ddots & \\ & & \ddots & 1 \\ & & & 0 \end{bmatrix}.$$

Then $E^j = 0$ for $j \ge n$ where n is the dimension of E. Hence,

$$J_1^k = \begin{bmatrix} \lambda^k & & \\ & \ddots & \\ & & \lambda^k \end{bmatrix}, \quad J_2^k = [J_1 + E]^k = \sum_{j=0}^k \binom{k}{j} J_1^{k-j} E^j$$

$$= \sum_{j=0}^{n-1} \binom{k}{j} J_1^{k-j} E^j\ .$$

there exists a positive constant K such that whenever $x_0=0$ and $|u_k|\leq 1$ for all $k\geq 0$, we have $|v_k|\leq K$. If $\Sigma_{j=1}^k|h_j|$ is unbounded, then for each (arbitrarily large) positive constant N, we can choose $k_1>0$ such that $\Sigma_{j=1}^{k_1}|h_j|>pqN$. Hence, if we denote by h_{jlk} the (l, k)th entry of the $q\times p$ matrix h_j, then

$$pqN<\sum_{j=1}^{k_1}|h_j|\leq\sum_{j=1}^{k_1}\left(\sum_{l=1}^{q}\sum_{k=1}^{p}h_{jlk}^2\right)^{1/2}\leq\sum_{j=1}^{k_1}\sum_{l=1}^{q}\sum_{k=1}^{p}|h_{jlk}|\leq pq\sum_{j=1}^{k_1}|h_{j\alpha\beta}|$$

for some (α, β) where $1\leq\alpha\leq q$ and $1\leq\beta\leq p$. That is, $\Sigma_{j=1}^{k_1}|h_{j\alpha\beta}|>N$. Let

$$u_{k_1-j}=[0\ldots 0\ \text{sgn}\{h_{j\alpha\beta}\}\ 0\ldots 0]^T\ ,$$

where $\text{sgn}\{h_{j\alpha\beta}\}$ is placed at the βth component of u_{k_1-j}. Then

$$|v_{k_1}|^2=\left|\sum_{l=0}^{k_1-1}h_{k-l}u_l\right|^2=\left|\sum_{j=1}^{k_1}h_ju_{k_1-j}\right|^2\geq\left(\sum_{j=1}^{k_1}|h_{j\alpha\beta}|\right)^2>N^2\ ,$$

a contradiction.

6.20 $zX(z)=AX(z)+BU(z)$ and $V(z)=CX(z),$

$$V(z)=CX(z)=C[zI-A]^{-1}BU(z)=\frac{C(zI-A)^*B}{\det(zI-A)}\ .$$

6.21 Let r be the radius of convergence of $\Sigma_0^\infty a_nw^n$. If $r>1$, then $\overline{\lim}_{n\to\infty}|a_n|^{1/n}=1/r<1$ so that $\Sigma_0^\infty|a_n|<\infty$. Conversely, if $\Sigma_0^\infty|a_n|<\infty$, then $|a_n|\to 0$ as $n\to\infty$. Hence, $r=\lim|a_n|^{-1/n}\geq 1$. Since $f(w)$ is a rational function, $f(w)$ has only finitely many poles, say at $z_1,\ldots,z_n$, and $|z_k|\geq 1$ for all k. We will see that $|z_k|>1$ for all k. Suppose $|z_1|=1$. Then, rewriting $f(z)$ as

$$f(w)=p(w)+\left(\frac{b_{11}}{w-z_1}+\ldots+\frac{b_{1m_1}}{(w-z_1)^{m_1}}\right)+\ldots$$

$$+\left(\frac{b_{n1}}{w-z_n}+\ldots+\frac{b_{n,m_n}}{(w-z_n)^{m_n}}\right),$$

where p is a polynomial and $1\leq m_i<\infty$, $i=1,\ldots,n$, it follows from

$$\frac{1}{w-z_1}=-\frac{1}{z_1}\frac{1}{1-\dfrac{w}{z_1}}=-\frac{1}{z_1}\sum_{n=0}^{\infty}\left(\frac{1}{z_1}\right)^n w^n\ ,$$

that if $m_1=1$, we have

$$\sum_0^\infty|a_n|\geq\text{const}+\sum_0^\infty\frac{1}{|z_1|^{n+1}}=\infty\ .$$

If $m_1 > 1$, we can prove the same result by induction. Hence $|z_k| > 1$ for all k. It follows that $f(w)$ is analytic in $|w| < r$ where $r > 1$.

6.22

$$A = \begin{bmatrix} 0 & 1 \\ 1 & 0 \end{bmatrix}, \quad B = \begin{bmatrix} 0 \\ 1 \end{bmatrix}, \quad \text{and} \quad C = [-1 \quad 1] \ .$$

This system is completely controllable, and since one of the eigenvalues of A is 1, the system is not asymptotically stable. Since $H(s) = 1/(s+1)$, the system is I−O stable.

6.23

$$A = \begin{bmatrix} 0 & 1 \\ 1 & 0 \end{bmatrix}, \quad B = \begin{bmatrix} -1 \\ 1 \end{bmatrix}, \quad \text{and} \quad C = [0 \quad 1] \ .$$

Chapter 7

7.1 Let $x_1 = \theta$, $x_2 = \dot{\theta}$. Then

$$\underset{|u| \le 1}{\text{minimize } F(u)}: F(u) = \int_0^{t_1} 1 \, dt \ ,$$

$$\begin{bmatrix} \dot{x}_1 \\ \dot{x}_2 \end{bmatrix} = \begin{bmatrix} 0 & 1 \\ -w_0^2 & -a \end{bmatrix} \begin{bmatrix} x_1 \\ x_2 \end{bmatrix} + \begin{bmatrix} 0 \\ 1 \end{bmatrix} u \ ,$$

$$\begin{bmatrix} x_1(0) \\ x_2(0) \end{bmatrix} = \begin{bmatrix} \theta_0 \\ \theta_1 \end{bmatrix}, \quad \begin{bmatrix} x_1(t_1) \\ x_2(t_1) \end{bmatrix} = \begin{bmatrix} 0 \\ 0 \end{bmatrix} \ .$$

7.2 A Bolza problem can be reformulated as a Mayer problem by adding an extra coordinate x_{n+1} and using the Pontryagin function with $F(u) = h(t_1, x(t_1)) + [0 \ldots 0 \ 1] \begin{bmatrix} x^{(t_1)} \\ x_{n+1}(t_1) \end{bmatrix}$. A Mayer problem can be changed to a Lagrange problem by letting $F(u) = h(t_1, x(t_1)) = \int_{t_0}^{t_1} [h(t_1, x(t_1))/(t_1 - t_0)] \, dt$. A Lagrange problem can be converted to a Bolza problem by simply choosing $h = 0$.

7.3 Suppose that $k_i(t)$, the ith component of $k(t)$, is not zero at $t = t_2 \in [t_0, t_1]$. Without loss of generality, suppose $k_i(t_2) > 0$. Then by the continuity of $k_i(t)$, there exists a neighborhood $N(t_2, \delta)$ of t_2 on which $k_i(t) > 0$. Choose $\eta(t) = [0 \ldots 0 \ \eta_i(t) \ 0 \ldots 0]^T$ where the ith component $\eta_i(t) > 0$ on $N(t_2, \delta)$. Then we have $\int_{t_0}^{t_1} k^T(t) \eta(t) \, dt > 0$, a contradiction.

7.4 $\dot{\xi} = \dfrac{d}{dt}(\delta x) = \dfrac{d}{dt} \lim_{\varepsilon \to 0} \dfrac{1}{\varepsilon} [x(u + \varepsilon \eta, t) - x(u, t)]$

$$= \lim_{\varepsilon \to 0} \frac{1}{\varepsilon} [\dot{x}(u + \varepsilon \eta, t) - \dot{x}(u, t)]$$

$$= \lim_{\varepsilon \to 0} \frac{1}{\varepsilon} [f(x(u + \varepsilon \eta, t), u + \varepsilon \eta, t) - f(x(u, t), u, t)]$$

$$= \lim_{\varepsilon \to 0} \frac{1}{\varepsilon} \{ f(x(u, t), u, t) + \frac{\partial f}{\partial x}(x(u, t), u, t)[x(u + \varepsilon \eta, t) - x(u, t)]$$

$$+ \frac{\partial f}{\partial u}(x(u, t), u, t)\varepsilon \eta + o(\varepsilon) - f(x(u, t), u, t) \}$$

$$= \frac{\partial f}{\partial x}(x, u, t)\xi + \frac{\partial f}{\partial u}(x, u, t)\eta .$$

7.5 $$0 = \delta_\eta F(u^*) = \int_{t_0}^{t_1} \left[\frac{\partial g}{\partial x}(x^*, u^*, t)\xi(t) + \frac{\partial g}{\partial u}(x^*, u^*, t)\eta(t) \right] dt$$

$$= \int_{t_0}^{t_1} \int_{t_0}^{t} \frac{\partial g}{\partial x}(x^*, u^*, t)\Phi(t, \tau)\frac{\partial f}{\partial u}(x^*, u^*, \tau)\eta(\tau)d\tau \, dt + \int_{t_0}^{t_1} \frac{\partial g}{\partial u}(x^*, u^*, t)\eta(t)dt$$

$$= \int_{t_0}^{t_1} \int_{\tau}^{t_1} \frac{\partial g}{\partial x}(x^*, u^*, t)\Phi(t, \tau)\frac{\partial f}{\partial u}(x^*, u^*, \tau)\eta(\tau)dt \, d\tau + \int_{t_0}^{t_1} \frac{\partial g}{\partial u}(x^*, u^*, \tau)\eta(\tau)dt$$

$$= \int_{t_0}^{t_1} \left[\int_{\tau}^{t_1} \frac{\partial g}{\partial x}(x^*, u^*, t)\Phi(t, \tau)\frac{\partial f}{\partial u}(x^*, u^*, \tau)dt + \frac{\partial g}{\partial u}(x^*, u^*, \tau) \right] \eta(\tau)dt .$$

The completeness of U implies (8.10)

7.6 Since $\dot{x}^* = x^* - p^*$ and $\dot{p}^* = -x^* - p^*$, we have $p^* = -\dot{x}^* - \dot{p}^* = -x^*$
$+ p^* - \dot{p}^* = 2p^*$. Hence, $p^*(t) = C_1 \exp(\sqrt{2}t) + C_2 \exp(-\sqrt{2}t)$. The two
boundary value conditions

$$p^*(1) = C_1 e^{\sqrt{2}} + C_2 e^{-\sqrt{2}} = 0$$

$$x^*(0) = -(\dot{p}^* + p^*)(0) = -(1 + \sqrt{2})C_1 - (1 - \sqrt{2})C_2 = 1$$

give

$$C_1 = \frac{-1}{(\sqrt{2} + 1) + (\sqrt{2} - 1)} \exp(2\sqrt{2}) \text{ and}$$

$$C_2 = \frac{1}{(\sqrt{2} - 1) + (\sqrt{2} + 1)} \exp(-2\sqrt{2}) .$$

7.7 Let $H = \frac{1}{2}[(x-1)^2 + u^2] + p(-x+u)$. Then $(\partial H / \partial u) = u + p = 0$ implies that $u^* = -p^*$. The two-point boundary value problem is

$$\begin{bmatrix} \dot{x}^* \\ \dot{p}^* \end{bmatrix} = \begin{bmatrix} -1 & -1 \\ -1 & 1 \end{bmatrix} \begin{bmatrix} x^* \\ p^* \end{bmatrix} + \begin{bmatrix} 0 \\ 1 \end{bmatrix},$$

$x^*(0) = 0$, $p^*(1) = 0$.

Finally,

$$u^* = \tfrac{1}{2} - \tfrac{1}{4}(1 + \sqrt{2})e^{\sqrt{2}t} - \tfrac{1}{4}(1 - \sqrt{2})e^{-\sqrt{2}t} .$$

7.8 Since the Hamiltonian is

$$H = \tfrac{1}{2}[x^T(t)Q(t)x(t) + u^T(t)R(t)u(t)] + p^T(t)[A(t)x(t) + B(t)u(t)] ,$$

we have, from (7.13), $u^*(t) = -R^{-1}(t)B^T(t)p(t)$ and hence

$$\dot{p}(t) = -A^T(t)p(t) - Q(t)x(t)$$

$$p(t_1) = 0 .$$

Let $p(t) = L(t)x(t)$. Then $L(t_1) = 0$ and for any nonzero $x(t)$ (determined by the arbitrarily given x_0), from the costate equation we have

$$[\dot{L}(t) + L(t)A(t) + A^T(t)L(t) - L(t)B(t)R^{-1}(t)B^T(t)L(t) + Q(t)]x(t) = 0 .$$

7.9 Since the Hamiltonian is

$$H = \tfrac{1}{2}[(y-v)^T Q(t)(y-v) + u^T R(t)u] + p^T[A(t)x + B(t)u] ,$$

from (7.13) it follows that $u^* = -R^{-1}B^T(t)p$ so that

$$\dot{p} = -A^T(t)p - C^T(t)Q(t)(y-v)$$

$$p(t_1) = 0 .$$

Let $p(t) = L(t)x - z$. Then for any nonzero x (determined by the arbitrarily given x_0), from the costate equation we have

$$[\dot{L}(t) + L(t)A(t) + A^T(t)L(t) - L(t)B(t)R^{-1}(t)B^T(t)L(t)$$

$$+ C^T(t)Q(t)C(t)]x + \{\dot{z} + [A(t) - B(t)R^{-1}(t)B^T(t)L(t)]z$$

$$+ C^T(t)Q(t)y\} = 0 ,$$

$$L(t_1)x(t_1) - z(t_1) = 0 .$$

7.10 From (7.5) we have

$$\delta u_k = \delta_{\eta_k} u_k = \lim_{\varepsilon \to 0} \frac{1}{\varepsilon}[u_k + \varepsilon \eta_k - u_k] = \eta_k, \quad \text{and}$$

$$\delta x_k = \lim_{\varepsilon \to 0} \frac{1}{\varepsilon} \left[x_k(u_k + \varepsilon \eta_k) - x_k(u_k) \right] = \frac{\partial x_k}{\partial u} \eta_k .$$

For convenience, we will simply write x_k, u_k instead of x_k^*, u_k^*, respectively. A necessary condition is $\delta F = 0$, i.e.

$$\delta F = \sum_{k=k_0}^{k_1} \left[\frac{\partial g}{\partial x}(x_k, u_k, k) \delta x_k + \frac{\partial g}{\partial u}(x_k, u_k, k) \delta u_k \right] = 0 .$$

Since $x_{k+1} = f(x_k, u_k, k)$, it follows that

$$\delta x_{k+1} = \frac{\partial f}{\partial x}(x_k, u_k, k) \delta x_k + \frac{\partial f}{\partial u}(x_k, u_k, k) \delta u_k, \quad k = k_0, k_0 + 1, \ldots, k_1 - 1 ,$$

$$\delta x_{k_0} = 0 . \tag{1}$$

Let Φ_{kj}, $k \geq j$, be the transition matrix of (1). Then

$$\delta x_k = \sum_{j=k_0+1}^{k} \Phi_{kj} \frac{\partial f}{\partial u}(x_{j-1}, u_{j-1}, j-1) \delta u_{j-1} , \tag{2}$$

$$k = k_0 + 1, k_0 + 2, \ldots, k_1 .$$

Substituting (2) into (1), we obtain

$$
\begin{aligned}
0 &= \sum_{k=k_0+1}^{k_1} \left[\frac{\partial g}{\partial x}(x_k, u_k, k) \sum_{j=k_0+1}^{k_1} \Phi_{kj} \frac{\partial f}{\partial u}(x_{j-1}, u_{j-1}, j-1) \delta u_{j-1} \right. \\
&\quad \left. + \frac{\partial g}{\partial u}(x_k, u_k, k) \delta u_k \right] + \frac{\partial g}{\partial u}(x_{k_0}, u_{k_0}, k_0) \delta u_{k_0} \\
&= \sum_{k=k_0+1}^{k_1} \sum_{j=k_0+1}^{k_1} \frac{\partial g}{\partial x}(x_k, u_k, k) \Phi_{kj} \frac{\partial f}{\partial u}(x_{j-1}, u_{j-1}, j-1) \delta u_{j-1} \\
&\quad + \sum_{k=k_0}^{k_1} \frac{\partial g}{\partial u}(x_k, u_k, k) \delta u_k \\
&= \sum_{j=k_0+1}^{k_1} \left[\sum_{k=j}^{k_1} \frac{\partial g}{\partial x}(x_k, u_k, k) \Phi_{kj} \right] \frac{\partial f}{\partial u}(x_{j-1}, u_{j-1}, j-1) \delta u_{j-1} \\
&\quad + \sum_{j=k_0+1}^{k_1+1} \frac{\partial g}{\partial u}(x_{j-1}, u_{j-1}, j-1) \delta u_{j-1} \\
&= \sum_{j=k_0+1}^{k_1} \left\{ \left[\sum_{k=j}^{k_1} \frac{\partial g}{\partial x}(x_k, u_k, k) \Phi_{kj} \right] \frac{\partial f}{\partial u}(x_{j-1}, u_{j-1}, j-1) \right. \\
&\quad \left. + \frac{\partial g}{\partial u}(x_{j-1}, u_{j-1}, j-1) \right\} \delta u_{j-1} + \frac{\partial g}{\partial u}(x_{k_1}, u_{k_1}, k_1) \delta u_{k_1} .
\end{aligned}
$$

Since the sequence $\{\delta u_{k_0}, \delta u_{k_0+1}, \ldots, \delta u_{k_1-1}, \delta u_{k_1}\}$ can be arbitrarily chosen as long as it is in the admissible class which contains the "delta sequences", by choosing it to be $\{0, \ldots, 0, e_i\}$, $\{e_i, 0, \ldots, 0\}, \{0, e_i, \ldots, 0\}, \ldots, \{0, \ldots, e_i, 0\}$ respectively, where $e_i = [0 \ldots 0\, 1\, 0 \ldots 0]^T$ with 1 being placed at the ith component, $i = 1, \ldots p$, we obtain

$$\frac{\partial g}{\partial u}(x_{k_1}, u_{k_1}, k_1) = 0 \qquad \text{and} \tag{3}$$

$$\left[\sum_{k=j}^{k_1} \frac{\partial g}{\partial x}(x_k, u_k, k)\Phi_{kj}\right]\frac{\partial f}{\partial u}(x_{j-1}, u_{j-1}, j-1) + \frac{\partial g}{\partial u}(x_{j-1}, u_{j-1}, j-1) = 0 \,,$$

$$j = k_0 + 1, \ldots, k_1 \,. \tag{4}$$

To simplify (4), define the costate p_k to be the unique solution of

$$p_k = \left[\frac{\partial f}{\partial x}(x_k, u_k, k)\right]^T p_{k+1} + \left[\frac{\partial g}{\partial x}(x_k, u_k, k)\right]^T, \quad k = k_1, k_1 - 1, \ldots, k_0 \,,$$

$$p_{k_1+1} = \overset{\cdot}{0} \,,$$

and denote $A_k = \frac{\partial f}{\partial x}(x_k, u_k, k)$ and $b_k^T = \frac{\partial g}{\partial x}(x_k, u_k, k)$. Then

$$p_{k_1} = b_{k_1}, \; p_{k_1-1} = A_{k_1-1}^T p_{k_1} + b_{k_1-1}, \ldots ,$$

$$p_j = A_j^T A_{j+1}^T \ldots A_{k_1-1}^T b_{k_1} + A_j^T A_{j+1}^T \ldots A_{k_1-2}^T b_{k_1-1}$$

$$+ \ldots + A_j^T b_{j+1} + b_j$$

$$= \sum_{k=j}^{k_1} \Phi_{kj}^T \left[\frac{\partial g}{\partial x}(x_k, u_k, k)\right]^T, \quad j = k_1, \ldots, k_0 \,,$$

where $\Phi_{kj} = A_{k-1} \ldots A_j$ is the transition matrix of (1). Hence, (3) and (4) can be rewritten as

$$p_j^T \frac{\partial f}{\partial u}(x_{j-1}, u_{j-1}, j-1) + \frac{\partial g}{\partial u}(x_{j-1}, u_{j-1}, j-1) = 0 \,,$$

$$j = k_0 + 1, \ldots, k_1 \,. \tag{5}$$

Furthermore, if we define the Hamiltonian to be

$$H(x_k, u_k, p_{k+1}, k) = g(x_k, u_k, k) + p_{k+1}^T f(x_k, u_k, k) \,,$$

then (5) is equivalent to

$$\frac{\partial H}{\partial u}(x_k, u_k, p_{k+1}, k) = 0, \; k = k_0 + 1, \ldots, k_1 \,.$$

7.11   Since the Hamiltonian is

$$H = \tfrac{1}{2}(x_k^T Q_k x_k + u_k^T R_k u_k) + p_{k+1}^T (A_k x_k + B_k u_k) \ ,$$

we have, from Theorem 7.2, $u_k^* = -R_k^{-1} B_k^T p_{k+1}$ and hence,

$$p_k = A_k^T p_{k+1} + Q_k x_k, \qquad k = k_1, \ldots, k_0 \ ,$$

$$p_{k_1 + 1} = 0 \ .$$

Let $p_k = L_k x_{k-1}$, $k = k_1, \ldots, k_0 + 1$. Then for any nonzero x_{k-1} (determined by the arbitrarily given x_{k_0}) we have $L_{k_1 + 1} = 0$ and from the costate equation, we have

$$[L_k - A_k^T L_{k+1} A_{k-1} + Q_k B_{k-1} R_{k-1}^{-1} L_k + A_k^T L_{k+1} B_{k-1} R_{k-1}^{-1} B_{k-1}^T L_k$$
$$- Q_k A_{k-1}] x_{k-1} = 0 \ .$$

Chapter 8

8.1   Since the Hamiltonian is $H = \tfrac{1}{2}(x^2 + u^2) + pu$, $(\partial H/\partial u) = u + p$ so that $u^* = -p^*$. Solving the two-point boundary value problem

$$\dot{x}^* = -p^*, \quad \dot{p}^* = -x^*$$

$$x(0) = 1, \quad p^*(2) = 0 \ ,$$

we obtain

$$x^*(t) = \frac{e^{-2}}{e^2 + e^{-2}} e^t + \frac{e^2}{e^2 + e^{-2}} e^{-t} \quad \text{and}$$

$$u^*(t) = \frac{e^{-2}}{e^2 + e^{-2}} e^t - \frac{e^2}{e^2 + e^{-2}} e^{-t} \ .$$

Solving the second problem, we obtain the same optimal control u^*.

8.2 $$\min_{u \in U} \left\{ \int_t^\tau g(x, u, s)\, ds + \int_\tau^{t_1} g(x, u, s)\, ds \right\}$$

$$\geq \min_{u \in U} \left\{ \int_t^\tau g(x, u, s)\, ds + \min_{\tilde{u} \in U} \int_\tau^{\tilde{t}_1} g(\tilde{x}, \tilde{u}, s)\, ds \right\}$$

$$= \min_{u \in U} \left\{ \int_t^\tau g(x, u, s)\, ds + \int_\tau^{\tilde{t}_1} g(\tilde{x}, \tilde{u}, s)\, ds \right\} \quad [\text{for some } (\tilde{u}, \tilde{x}) \text{ and } \tilde{t}_1]$$

$$= \int\limits_{t}^{\tau} g(\hat{x}, \hat{u}, s)\, ds + \int\limits_{\tau}^{\tilde{t}_1} g(\tilde{x}, \tilde{u}, s)\, ds \quad [\text{for some } (\hat{u}, \hat{x})]$$

$$= \int\limits_{t}^{\tilde{t}_1} g(\bar{x}, \bar{u}, s)\, ds$$

$$\geq \min_{u \in U} \left\{ \int\limits_{t}^{\tilde{t}_1} g(x, u, s)\, ds \right\}$$

$$= \min_{u \in U} \left\{ \int\limits_{t}^{\tau} g(x, u, s)\, ds + \int\limits_{\tau}^{t_1} g(x, u, s)\, ds \right\}, \quad \text{where}$$

$$g(\bar{x}, \bar{u}, s) = \begin{cases} g(\hat{x}, \hat{u}, s), & t \leq s \leq \tau, \\ g(\tilde{x}, \tilde{u}, s), & \tau \leq s \leq \tilde{t}_1. \end{cases}$$

8.3 Solving the minimization problem (8.4), i.e.

$$\min_{u \in U} \left\{ \tfrac{1}{2}[x^T Q x + u^T R u] + \left[\frac{\partial V}{\partial x}\right](Ax + Bu) \right\},$$

we obtain

$$u^* = -R^{-1} B^T \left[\frac{\partial V}{\partial x}\right]^T.$$

Substituting u^* and the linear system equation into (8.3), we arrive at the required form. For any nonzero x (determined by the arbitrarily given x_0), let $V = \tfrac{1}{2} x^T L(t) x$. Then $L(t_1) = 0$ and

$$\tfrac{1}{2} x^T [\dot{L} + LA + A^T L - LBR^{-1} B^T L + Q] x = 0.$$

8.4 Since $u^* = -\partial V/\partial x = a(t)x$ so that

$$\dot{x}^* = [1 + a(t)] x^* = \frac{\dot{b}(t)}{b(t)} x^*$$

$$x^*(0) = 1, \quad \text{where}$$

$$a(t) = \frac{e^{-\sqrt{2}(1-t)} - e^{\sqrt{2}(1-t)}}{(\sqrt{2}+1)e^{-\sqrt{2}(1-t)} + (\sqrt{2}-1)e^{\sqrt{2}(1-t)}} \quad \text{and}$$

$$b(t) = (\sqrt{2}+1)e^{-\sqrt{2}(1-t)} + (\sqrt{2}-1)e^{\sqrt{2}(1-t)},$$

we have

$$x^*(t) = \frac{(\sqrt{2}+1)e^{-\sqrt{2}(1-t)} + (\sqrt{2}-1)e^{\sqrt{2}(1-t)}}{(\sqrt{2}+1)e^{-\sqrt{2}} + (\sqrt{2}-1)e^{\sqrt{2}}}$$

and hence

$$u^*(t) = \frac{e^{-\sqrt{2}(1-t)} - e^{\sqrt{2}(1-t)}}{(\sqrt{2}+1)e^{-\sqrt{2}} + (\sqrt{2}-1)e^{\sqrt{2}}}$$

$$= \frac{\sqrt{2}-1}{(3-2\sqrt{2})e^{2\sqrt{2}}+1}e^{\sqrt{2}t} - \frac{\sqrt{2}+1}{(3+2\sqrt{2})e^{-2\sqrt{2}}+1}e^{-\sqrt{2}t} .$$

8.5 $\lambda = -a^{-1}$.

8.6 Substituting $x = 1/z + x_1$ into Riccati's equation, we have

$$\dot{z} + [b(t) + 2a(t)x_1]z + a(t) + [-\dot{x}_1 + a(t)x_1^2 + b(t)x_1 + c(t)]z^2 = 0 .$$

Since x_1 is a particular solution of the Riccati equation, the coefficient of z^2 is zero.

8.7 Imitate the procedure used in solving the one-dimensional example in this section.

8.8 Lemma 8.3 can be proved by imitating the procedure used in proving Lemma 8.2 (see the answer to Exercise 8.2). Theorem 8.2 can be proved by using Lemma 8.3 repeatedly.

8.9 Let V_n be the minimum value of the sum $\Sigma_{i=1}^n r_i$. By Lemma 8.3 we have

$$V_n = \min_{0 \le r_1 \le r} \left\{ r_1 + V_{n-1}\left(\frac{r}{r_1}\right) \right\}, \; n \ge 2 .$$

Since $V_1(r) = r$, $V_1(r/r_1) = r/r_1$. Hence, when $n=2$,

$$V_2(r) = \min_{0 \le r_1 \le r} \left\{ r_1 + \frac{r}{r_1} \right\} .$$

Using calculus, we obtain $r_1^* = \sqrt{r}$ so that $r_2^* = \sqrt{r}$ and $V_2 = 2\sqrt{r}$. When $n=3$,

$$V_3(r) = \min_{0 \le r_1 \le r} \left[r_1 + V_2\left(\frac{r}{r_1}\right) \right] = \min_{0 \le r_1 \le r} \left[r_1 + 2\left(\frac{r}{r_1}\right)^{1/2} \right] .$$

Using calculus agian, we obtain $r_1^* = r^{1/3}$ and so $V_2(r/r_1^*) = r^{2/3}$. Minimizing this V_2 by using the above procedure, we have $r_2^* = r_3^* = r^{1/3}$. By induction, we obtain $r_i^* = r^{1/n}$, $i = 1, \ldots, n$.

8.10 If the terminal time t_1 is fixed, we have the same two-point boundary value problems as those in Exercises 7.7–9.

8.11 From Theorem 8.3, we have $\dot{p} = p$ with $p(1) = 0$ which implies that $p = 0$. Hence we need to find u^* such that

$$|u^*| = \min_{u \in U}|u|$$

subject to $\dot{x}^* = x^* + u^*$, $x^*(0) = 0$ and $x^*(1) = 1$. From

$$1 = x^*(1) = \int_0^1 e^{1-t} u^* \, dt = (e-1)u^* \ ,$$

we obtain $u^* = 1/(e-1)$.

Chapter 9

9.1 Without loss of generality, suppose that $y_i(t)$, the ith component of $y(t)$, is positive on some subset E with positive measure in $[t_0, t_1^*]$ and that $u_i^*(t) \neq \text{sgn}\{y_i(t)\}$ on E. Then $u_i^*(t) < 1 - \varepsilon$ for some $\varepsilon > 0$ on E. Define $\hat{u}(t) = u^*(t)$ except that $\hat{u}_i(t) = 1$ on E. Then we have $\hat{u} \in W$ and

$$y^T(t)\hat{u}(t) > y^T(t)u^*(t) = \max_{u \in W} y^T(t)u(t) \ ,$$

a contradiction.

9.2 $u^*(t) = \begin{cases} -1, & 0 \leq t < 1 + \frac{1}{2}\sqrt{22}, \\ 1, & \frac{1}{2}\sqrt{22} \leq t \leq t_1^* = 3 + \sqrt{22} \ . \end{cases}$

9.3 From (9.9) we have $q_2(t) = -e^{at/2}(z_1 t + z_2)$ where $q(t) = [q_1(t) \ q_2(t)]^T$ and $z = [z_1 \ z_2]^T$ so that $u^* = -\text{sgn}\{B^T q(t)\} = \text{sgn}\{e^{at/2}(z_1 t + z_2)\}$.

9.4 Since

$$1 = x(t_1) = \int_{t_0}^{t_1} [u(s) - u^2(s)] \, ds \leq (t_1 - t_0)\max(u - u^2) \ ,$$

and $(u - u^2)$ assumes its maximum at $u = \frac{1}{2}$, we have $t_1^* - t_0 = 4$ and $u^* \equiv 1/2$.

9.5 $M_{AB} = \begin{bmatrix} 0 & 1 \\ 1 & 0 \end{bmatrix}$ is of full rank and has eigenvalues $\lambda_1 = 1$ and $\lambda_2 = -1$.

9.6 M_{AB} is of full rank, hence the system is normal. The optimal control function is $u^*(t) = \text{sgn}\{\frac{1}{2}t^2 z_1 - t z_2 + z_3\}$.

9.7 Writing $B = [b_1 \ \ldots \ b_p]$ and observing $u_i^* = \text{sgn}\{z^T \exp[-(t - t_0)A]b_i\}^T$, $i = 1, \ldots, p$, we can prove the result for each i, $i = 1, \ldots, p$, by imitating the proof of Theorem 9.5.

9.8 $e^{-(t-t_0)A} = I - (t - t_0)A + \dfrac{(t - t_0)^2}{2!}A^2 - \dfrac{(t - t_0)^3}{3!}A^3 + \ldots$

$$= I - (t - t_0)P \, \text{diag}[\lambda_1, \ldots, \lambda_n]P^{-1}$$

$$+ \frac{(t - t_0)^2}{2!}P \, \text{diag}[\lambda_1^2, \ldots, \lambda_n^2]P^{-1} - \ldots$$

$$= P\left\{I-(t-t_0)\operatorname{diag}[\lambda_1,\ldots,\lambda_n]\right.$$

$$\left.+\frac{(t-t_0)^2}{2!}\operatorname{diag}[\lambda_1^2,\ldots,\lambda_n^2]-\ldots\right\}P^{-1}$$

$$= P\operatorname{diag}[e^{-\lambda_1(t-t_0)},\ldots,e^{-\lambda_n(t-t_0)}]P^{-1}\ .$$

9.9 When $k=1$, $c_1(t)\exp[\mu_1(t-t_0)]$ has the same zeros as $c_1(t)$ and hence has at most m_1-1 positive zeros. Assume that $h_{k-1}(t)$ has at most $m_1+\ldots+m_{k-1}-1$ positive zeros but $h_k(t)$ has at least $m_1+\ldots+m_{k-1}+m_k$ positive zeros. Then

$$e^{-\mu_k(t-t_0)}h_k(t)=\sum_{j=1}^{k}c_j(t)e^{(\mu_j-\mu_k)(t-t_0)}$$

has also at least $m_1+\ldots+m_k$ positive zeros. Hence, the m_kth derivative of $\exp[-\mu_k(t-t_0)]h_k(t)$, which is $\Sigma_{j=1}^{k-1}\tilde{c}_j(t)\exp[(\mu_j-\mu_k)(t-t_0)]$ where $(\mu_j-\mu_k)$ are distinct and $\tilde{c}_j(t)$ is a polynomial of degree m_j-1 for each j, has at least $m_1+\ldots+m_k-m_k=m_1+\ldots+m_{k-1}$ positive zeros. This contradicts the induction hypothesis.

Notation

x, $x(t)$, x_k $n \times 1$ state vectors
u, $u(t)$, u_k $p \times 1$ vector-valued control (or input) functions, $p \leq n$
v, $v(t)$, v_k $q \times 1$ vector-valued output functions, $q \leq n$
x_e equilibrium point (or state)   49
x^* optimal trajectory (or state)   72
$\{x_k^*\}$ optimal trajectory (or state) sequence   78
u^* optimal control function   72
$\{u_k^*\}$ optimal control sequence   78
u_{bb}^* optimal (bang-bang) control function   98
x_0, x_{k_0} initial states   9, 78
x_1, x_{k_1} target positions   87, 94
t_1, k_1 terminal times   70, 78
t_1^*, k_1^* optimal terminal times   83, 87
p costate   74
p^* optimal costate   74
$\{p_k^*\}$ optimal costate sequence   79
X subset in $\mathbb{R}^n$ to which all trajectories are confined   70, 81
X_T losed subset of X 81
$\mathscr{U}$, U, W admissible classes of control functions   14, 70, 94
$U(\tau, y)$ subset of admissible control functions   81
W_{bb} set of bang-bang control functions   96
J time interval   8, 70
J_T closed sub-interval of J 81
M_T target, $M_T = J_T \times X_T$ 81

$A(t)$, A $n \times n$ system (or dynamic) matrices
$B(t)$, B $n \times p$ control matrices, $p \leq n$
$C(t)$, C $q \times n$ observation (or output) matrices, $q \leq n$
$D(t)$, D $q \times p$ transfer matrices
$\Phi(t, \tau)$, Φ_{ij} state transition matrices   8, 13
$G(t)$ gain matrix   108
$H(x, u, p, t)$, $H(x_k, u_k, p_{k+1}, k)$ Hamiltonians   75, 78
J Jordan canonical form   58
$\det A$ determinant of matrix A
A^* adjoint matrix of A 44

$\|A\|$ operator norm of matrix A, $\|A\| := \sup\{|Ax|_2 : |x|_2 = 1\}$ 60

$|A|_p$ l^p norm of matrix (or vector) A, $|A|_p := \left(\sum_{i,j} |a_{ij}|^p\right)^{1/p}$, $1 \leq p < \infty$

$|A|_\infty$ l^∞ norm of matrix (or vector) A, $|A|_\infty := \max_{i,j} |a_{ij}|$

$|A| := |A|_2$ 51, 96

$\mathrm{diag}[\lambda_1, \ldots, \lambda_n]$ diagonal matrix

$\mathscr{S}$ linear system

$\mathscr{S}_c$ continuous-time linear system

$\mathscr{S}_d$ discrete-time linear system

$\mathrm{sp}\{x_1, \ldots, x_n\}$ linear algebraic span of set $\{x_1, \ldots, x_n\}$ 14, 37

v null space of

$\oplus$ direct sum of

$L_\infty[t_0, t_1]$ space of almost everywhere bounded functions   95

$\mathscr{L}$ Laplace transform   43

Z z-transform   43

$H(s)$ transfer function   44

$H(z)$ transfer function   66

sgn signum function   63

$V(x, t)$ Lyapunov function   111

$V(\tau, y)$ value function   83

$q_m(s)$ minimum polynomial   45

$L_t u := \int_{t_0}^t \Phi(t, s) B(s) u(s)\, ds$ 18

$Q_t := \int_{t_0}^t \Phi(t, s) B(s) B^T(s) \Phi^T(t, s)\, ds$ 18

$P_t := \int_{t_0}^t \Phi^T(\tau, t_0) C^T(\tau) C(\tau) \Phi(\tau, t_0)\, d\tau$ 27

$R_{l^*} := \sum_{i=l+1}^{l^*} \Phi_{l^*i} B_{i-1} B_{i-1}^T \Phi_{l^*i}^T$ 22

$S_{l^*}\{u_k\} := \sum_{k=l+1}^{l^*} \Phi_{l^*k} B_{k-1} u_{k-1}$ 23

$L_m := \sum_{k=l+1}^{m} \Phi_{kl}^T C_k^T C_k \Phi_{kl}$ 30

$M_{AB} := [B\ AB\ \ldots\ A^{n-1}B]_{n \times pn}$, controllability matrix   20

$N_{CA} := \begin{bmatrix} C \\ CA \\ \vdots \\ CA^{n-1} \end{bmatrix}_{qn \times n}$, observability matrix   28

$T_{CA} := \begin{bmatrix} C \\ CA \end{bmatrix}$, total observability matrix   30

$h^*(t, s) := C(t)\Phi(t, s)B(s)$ 62

$h(t) := Ce^{tA}B$, impulse response   63

$h_j := CA^{j-1}B$, impulse response   65

$$R_t := \left\{ \int_{t_0}^{t} \Phi(t_0, s)B(s)u(s)ds : u \in W \right\}   95$$

X_T target set   81

$$X_t := \Phi(t, t_0)\{x_0 + R_t\} = \left\{ \Phi(t, t_0)x_0 + \int_{t_0}^{t} \Phi(t, s)B(s)u(s)ds : u \in W \right\}   95$$

$$K(u) := \int_{t_0}^{t} \Phi(t_0, s)B(s)u(s)ds   95$$

$$B_t := \left\{ \Phi(t, t_0)x_0 + \int_{t_0}^{t} \Phi(t, s)B(s)u(s)ds : u \in W_{bb} \right\} \subset X_t   97$$

$$V = V_y := \left\{ u \in W : y = \int_{t_0}^{t} \Phi(t_0, s)B(s)u(s)ds \right\}   97$$

$\partial R_{t_1^*}$ boundary of $R_{t_1^*}$ 98

δl variation of vector-valued function l 73

$\partial l / \partial u$ gradient of scalar-valued function l with respect to u, a row-vector   73

$\partial l / \partial u$ matrix, where both l and u are vectors   73

Subject Index